A DAINTREE DIARY

Tales from Travels to the Daintree Rainforest in tropical North Queensland, Australia

CARL PORTMAN

Typeset by Jonathan Downes,
Cover and Layout by SPiderKaT for CFZ Communications
Using Microsoft Word 2000, Microsoft , Publisher 2000, Adobe Photoshop CS.

First published in Great Britain by CFZ Press

**CFZ Press
Myrtle Cottage
Woolsery
Bideford
North Devon
EX39 5QR**

© CFZ MMX

All rights reserved. Without limiting the rights under copyright reserved above, no part of this publication may be reproduced, stored in or introduced into a retrieval system, or transmitted, in any form of by any means (electronic, mechanical, photocopying, recording or otherwise), without the prior written permission of both the copyright owners and the publishers of this book.

ISBN: 978-1-905723-53-9

Dedicated to

Steve Irwin

*May your spirit live on in the hearts and minds of
those that choose to walk but a single mile in your moccasins*

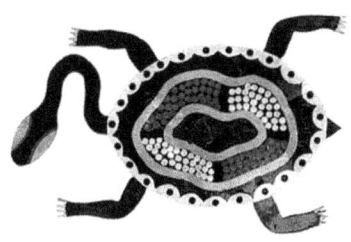

Acknowledgements

Many people have contributed, wittingly or unwittingly, to the making of this book. I especially want to thank my 'significant other' Susan Watson, whose support and encouragement was invaluable. Angela Leyshon is an integral part of this story, and I want to thank her for tolerating me, and for playing her part on an unforgettable trip. I am very appreciative of the people in Australia who gave of their time and knowledge to help me learn and develop my knowledge of animals in the wild; Neil and Prue at Cooper Creek for imparting their wisdom in such an open and generous manner; Clay Mitchell was very helpful, as were all the staff at the wonderful Mandalay Apartments in Port Douglas - Raelene and Nicole in particular.

Deepest thanks to Stuart Douglas at the Australian Venom Zoo, and to Mark and Anne Palfreyman for their wonderful supper and evening expedition. Further, I should like to thank Bill Clarke for doing his level best to help me catch a monster fish, and for being so open and illuminating during our fishing trip.

Tailor Made Travel – for their 'can do' approach and first class organisation. Thanks to Colin and Sheila Palmer for looking after our garden whilst we were away – the roses are lovely! I am also indebted to Anya and Nigel Young for looking after our Border Collie, Darwin, whilst we were away.

Getac computers trusted me with some seriously expensive equipment, for which they and Sarah Chard from the PR Room are to be applauded. Without past encouragement from the likes of Ray and Angela Hale, Mark Titterton, Ray Gabriel and many more at the British Tarantula Society I would never have developed my knowledge and love of arachnids. Thanks to "Geordie" and to Carl Marshall at the Stratford-Upon-Avon Butterfly Farm who allow me to hone my macro photography skills onsite, and are always very obliging.

Finally, without the support and belief from the incomparable Jon Downes I doubt that this book would have materialised. I am blessed to know such a man – thank you, Jon.

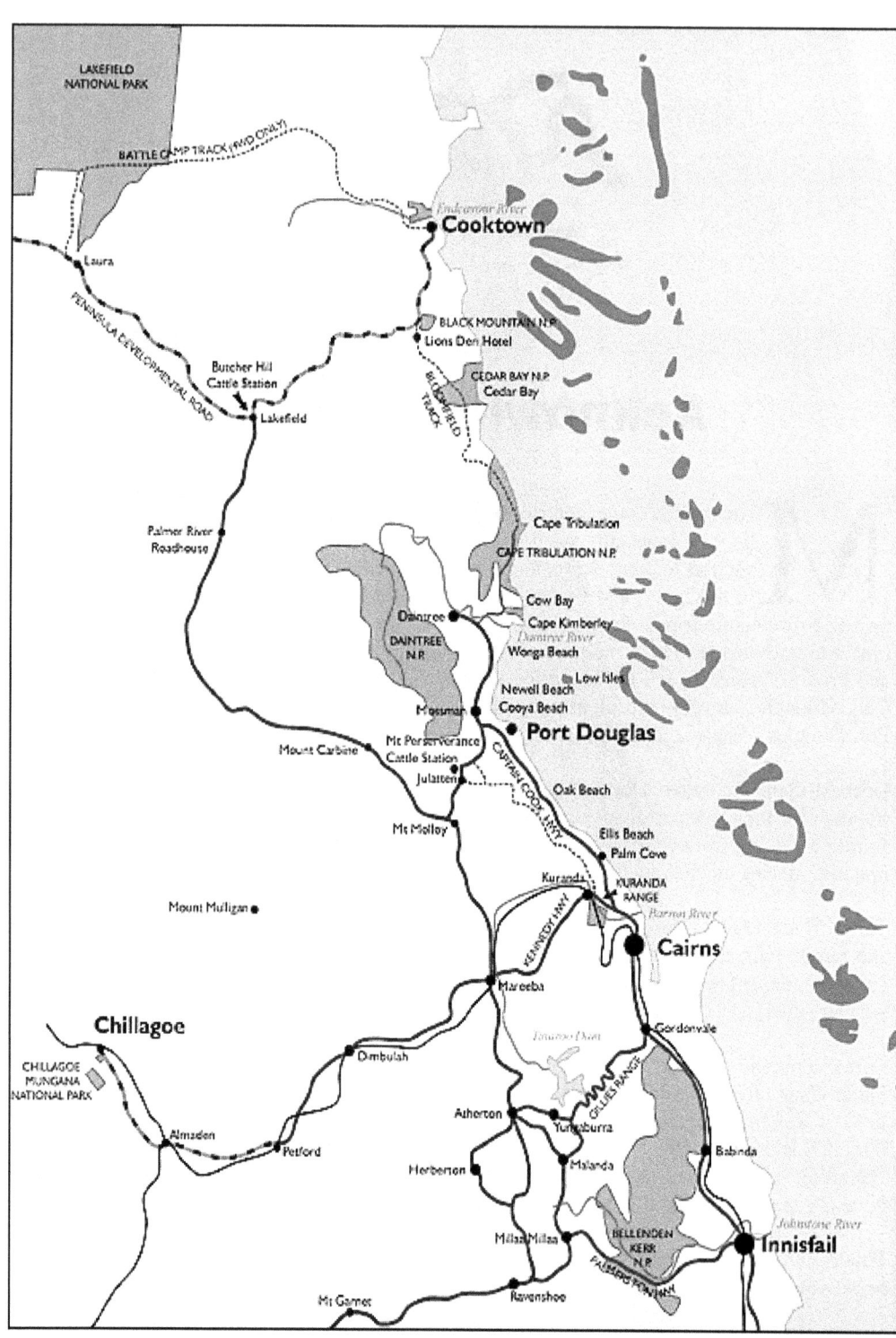

Contents

5.	Acknowledgements	
9.	Introduction	
11.	Chapter One.	Running out of planes
31.	Chapter Two.	Anyone for cane toad golf?
59.	Chapter Three.	Python patrol
87.	Chapter Four.	Spiders really do eat birds
93.	Chapter Five.	Above the canopy
115.	Chapter Six.	The jealous crocodile
131.	Chapter Seven.	The ghost in the forest
151.	Chapter Eight.	Down came a spider
161.	Chapter Nine.	The X-Factor debacle
175.	Chapter Ten.	The Blue Mountains
179.	Chapter Eleven.	Return to Londinium
183.	Chapter Twelve.	Post trip comments

WHAT IS MAN WITHOUT BEASTS?

If all the beasts were gone,
Man would die from great loneliness of spirit
For whatever happens to the beasts
Also happens to man
All things are connected.
Whatever befalls the earth
Befalls the sons of the earth

Chief Seattle of the Suquamish & Duwamish 1855

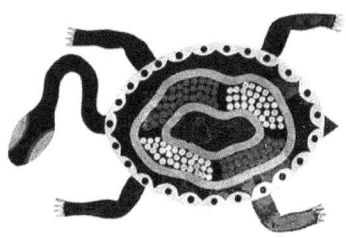

Introduction

This book is the diary of a short trip to Australia, which involved sacrifice and not a little courage. I hope it has something for everyone. Whilst the emphasis is on natural history, there's plenty of information about such subjects as trains, fear of flying, 'grumpy old man' moments, fascinating historical facts, Singapore and World War II, life, death, tears and laughter. There's poetry, quiz questions, bad language and even tales of dragons. What more could you possibly want?

Ever since I saw a programme about Far North Queensland by the late Steve Irwin I dreamed of leaving my own footprints on that unforgettable soil. I was gripped by the primitive landscape, and its wealth of unique flora and fauna. I was humbled at the age and size of the trees, and of the variety of plants and animals.

Australia not only has extraordinary animals today; it has always been home to curious beasts. I have been fascinated for many years; nay, almost fixated with an animal that is now sadly extinct. The creature was dog-like in appearance with stripes on its back like a tiger and a long, extraordinary tail. Only in my dreams can I actually visualise this magnificent animal running, hunting, sleeping, playing, fighting and nurturing its young. The thylacine was hunted to extinction and has joined the ranks of other animals whose passing we must now lament.

A pair of thylacines in Hobart Zoo prior to 1921.
The male in the background is much larger than the female. (Author of photo unknown)

I let the years go by without seriously asking myself what was stopping me going to Australia. I think as we get older we become more aware of our own mortality. When you are young it's money that matters but when you get older, it is time.

One day something within me said it was time and I began to think seriously about a trip but my fear of flying was a tremendous hurdle to overcome. It isn't that I am afraid to die in an aircraft, although the thought of being strapped into the seat of a plane smashed to bits whilst the skin burns off my body is pretty horrifying. It isn't going quickly that bothers me; it's the thought of surviving a crash in some sort of vegetative state. To add to this I suffer from terrible pain in my ears when flying and suffer crushing headaches. Nothing helps. My heart races at the slightest turbulence and I always seem to have to sit next to some sweaty behemoth with the communication skills of a lowland gorilla. If you hate flying too, I have news for you – you can overcome it just as I have. The only boundary is your mind. To get what you really want, especially if it is highly desirable, you have to make sacrifices. Nothing comes easy except to the blessed few. Now I have created those memories I want to share them with you. It is also for those many people who I know have similar noble ambitions but who for whatever reason fail to achieve them. Hopefully it will inspire them to turn 'I want to' into 'I did it'.

This trip involved three people: my partner Sue, her very good friend Angela and myself. We all shared council one evening over a glass of wine and formulated our strategy. The golden rule was that each of us had to get what we wanted from the holiday. Angela wanted the opportunity to relax on the beach, reading and absorbing the rays of a glorious antipodean sun. For Susan there was an element of that too, along with the opportunity to see the rainforest and most of all, visit old friends in Sydney. I wanted to indulge in my passion for natural history and photography. I openly admit that I am writing it from my own perspective and I know the girls will forgive me for that.

My particular love of arachnids and insects would be well rewarded in Australia, the land of dangerous spiders, snakes and many other weird and wonderful creatures. Watching it all on TV is one thing but there is no substitute for being there. I have studied and kept many thousands of tarantulas and scorpions, as well as snakes and other fascinating creatures such as amblypygids and solifugids. But now I would get the biggest kick from photographing them in their own backyard. We also wanted to gain an insight into the culture of Australia. I have met a handful of Aussies in my time and they were all fine people but meeting them on their own turf was an exciting idea. Like the Brits, they have a reputation: large hats replete with corks hanging from them; khaki shorts; XXXX lager; surfing; sheep-shearing; 'Sheilas'; 'G'day mate'. There had to be more to the Aussies than that. It was time to find out.

We agreed to set our base in Port Douglas in Far North Queensland and undertake our operations from there. The plan was to fly from England to Singapore for a few days, then fly on to Port Douglas for a good eighteen days before flying down to Sydney to see friends before returning to the UK via a very quick stop in Bangkok for refuelling.

This is an account of that memorable trip; a collection of memories if you will. In the words of P.D. McCormick, "*In joyful strains then let us sing Advance Australia Fair.*"

Chapter 1.
Running out of planes.

Sat 1st November

Departure for Singapore
In the end it was an unplanned event that motivated me most to step onto the aircraft: the good old British weather. When the taxi pulled up onto the drive at my house to take us to Heathrow airport it was cold, grey and raining hard. In addition, there was a biting wind and the sinister hint of snow amongst the frozen raindrops now stinging our faces. It never abated all the way to the airport.

England had just endured another summer that turned out to be a non-event and we were all in dire need of sunshine. More than three consecutive days would be a tremendous fillip for my thin, white frame. The truth is that I have turned into a grumpy old man in recent years. When I hit forty in 2004 I felt a sense of relief that this change in my character would be justified on the grounds that I was old enough to complain about almost anything. I was even old enough to take up bowls.

I admit that I was awfully tense; I swore I would never fly again, let alone to the other side of the world. The knowledge that I am getting onto a plane does strange things to me; I behave out of character, buying useless things at the airport just to pass the time and forget the nightmare to come. I always muse at the things they sell in the duty free. At that point in the time I do not want eau de cologne, a huge bar of Toblerone or a bloody wristwatch. I need lots of spare underwear, a substantial injection of anything this side of legal to calm me down and a written guarantee from someone of a much higher being that there won't be any turbulence and that we will get there safely.

Approaching security I had more reasons to moan when my very expensive laptop, loaned to me by GETAC Computers for the journey, dropped off the back of the gravity roller/carousel because the inept Frau on security managed to back up the bags and mine was the last one off the conveyor belt.

"Oh great," I said. "The bloody thing's broken before I even start".

Nobody seemed to be listening so I followed the throng through the processing plant that is security.

"Shoes," she said without a smile, "And your jacket, and your phone, and your bags."

Her colleague was masticating gum with the mandibular dexterity of a Friesian cow as she chipped in with "And empty your pockets too."

Then they gave each other that smart-arsed nod that Laurel and Hardy give to each other when they have done something to be proud of.

I emerged without being arrested or molested and felt quite relieved. We only had a couple of hours to go until flight time so I left my bags with the girls and browsed around the shops in the hope that I might buy something useful. I find it strangely mystifying that you have to surrender any liquids before you go through security then you can proceed to buy bottles of this and that and take them on the plane. A bottle can be an offensive weapon. So what if the liquid inside was purchased 'within' the duty free area?

I was feeling pretty upbeat about flying, actually. I had been preparing myself psychologically for weeks prior to these moments just as I would prepare for a chess game. I reminded myself that we were off to Singapore to stop over for a few days before heading to Darwin and Cairns. Hey, I can do this, I thought. The Imodium stayed in my bag (another victory for me) and I chose to take only Sudafed, which seems to help. Then the inevitable happened: the flight was delayed and the tension began to mount. Whilst Angela and Sue sat in the departure lounge I paced up and down, keeping my head together.

I walked past an unshaven gentleman reading *The Times* and the headline was 'The lumbering journey towards death - holiday jet terror' (referring to an incident that occurred in the last few days in Lanzarote that I hadn't wanted to know about).

Two hours later I was at the point of going home when we were told that we were free to board. I decided to pop to the loo just before we got on the cylinder of death but I really shouldn't have. Incredibly, *The Last Post* was being played through the loudspeaker system! I summoned all my positive thoughts and decided it was a cruel trick played by a greater being to test me. I had come too far to back down now. Checkmate was imminent.

It makes me laugh to think that we wait around airports for hours upon end and when it suits them airline staff suddenly want to rush everything and you're ushered through the sausage machine like lambs to the slaughter. I feel like saying "hold on a minute; I have been here for five hours so you can bloody well wait until I find my boarding pass!"

There was one final cruel twist prior to take-off. The Captain made a statement over the tannoy and I can actually quote him: *"Good evening, ladies and gentlemen. I apologise for the delay this evening. The only way I can put it is that we ran out of planes. We managed to get this one out of the maintenance workshops from Cardiff."*

I looked at Sue and said "Did he just say what I thought he said?"

Sue nodded and I recall murmuring something about how the captain can sometimes give you too much information.

The flight actually went very well. During the moments of turbulence I recall the words of Captain Keith Godfrey in his brilliant book *Flying Without Fear* stating that turbulence is not dangerous; it's just uncomfortable. It really helped me. I even managed to eat without wanting to projectile-vomit the lot over the back of the head of the person in the seat in front of me – which was nice.

The flight rolled into **Sunday 2nd November.**

Monday 3rd November

Singapore arrival

The body is a wonderful machine. When you are freezing in London, it shivers to help regulate temperature. When you next expose your frame to the open air and are blasted by humidity and very hot temperatures, it sweats to perform a similar job.

We were taken by taxi to the York Hotel, a pleasant dwelling on the famous Orchard Road. I can but agree with the comedian Lee Evans when he says you travel halfway around the world, hauling your baggage and then some chap steps forward as you arrive at the hotel and offers to carry it the last few yards! Personally I choose to carry it myself as a point of principle. The plan for the next couple of days was to mix shopping with sightseeing and a visit to Singapore Zoo.

Singapore is a fascinating country. Its very name means Lion Town (singa, 'lion'; pore, 'town'). It's a fact that the girls are generally slim and ridiculously pretty and the boys; well, they are boys. I was surprised to see a good proportion of youngsters smoking though thankfully it was prohibited indoors. I was only jealous, anyway; good luck to them.

It's a city of monetary fines. Eating or drinking on public transport will incur a fine. Selling chewing gum will incur a huge fine of several thousand dollars. Crossing the road in the wrong area will incur a fine. It's a very clean, tidy place that I found most agreeable. No styrofoam burger boxes on the floor, no vomit from the previous evening's soiree by inebriated Neanderthals and no discarded chewing gum to stick to your shoes.

Travel is truly broadening; travel is educational and travel is one of the great pleasures for human beings to experience. It allows us to put our own lives into some sort of perspective. Take a simple issue like a car, for example. In England anyone can buy a car but in Singapore this is not the case. There is a waiting list, and car tax seriously increases after ten years of use. Unsurprisingly, men love their cars here and they have the sort of climate to ensure they are nice and clean without having to slog through winter snow and ice. They love them so much that many men wash their cars in the morning and then take the bus to work. They care, the Singaporeans, and I am suitably impressed.

Singapore's motto is *Majulah Singapora* (Malay) meaning "Onward, Singapore." It has four main languages: Malay, English, Tamil and Mandarin. It is estimated that the population is not far off five million. There is, of course, a very British feel to the place. It has been under colonial rule. On 29th January 1819 Sir Thomas Stamford Raffles landed and considered Singapore to be a significant trading outpost. He signed a treaty with Sultan Hussein Shah on behalf of the British East India Company on 6th February 1819. It didn't take long to do business in those days! Southern Singapore was developed as a British trading post and settlement and from there it became a self-governing state from 1959, but still within the British Empire.

Of course, just prior to that was World War II. The British decided to build a naval base on the northern end of Singapore in response to the build up of Japanese forces. Britain wanted to protect its assets in southeast Asia but the warships and equipment were brought over to Europe as a result of the war with Germany. This left a gaping resource hole in Singapore.

During WWII the Imperial Japanese Army invaded Malaya, which culminated in the battle of Singapore. The British were simply unprepared for this as most of their forces were busy in Europe, and they were defeated in only six days, surrendering the supposedly impregnable fortress to General Tomoyuki Yamashita on 15th February 1942. The British Prime Minister, Sir Winston Churchill, described the surrender as *"The worst disaster and largest capitulation in British history."* Incredibly, though, the British re-focused and repossessed the island on 12th September 1945, a month after the Japanese surrender.

It is far too difficult to really appraise a place based on just a few days but I found the people and the city to be very hospitable. One of the things everyone should try to do is leisurely sip a 'Singapore Sling' at the famous Raffles hotel or at the very least, enjoy afternoon tea there.

I never got to indulge in such a memorable activity as I was too busy photographing the sights. Time was precious and I had no desire to sit and relax – it's not my style. We agreed that a half-day tour of Chinatown would be fun and we would decide what to do after that.

First, though, we unanimously agreed that a visit to the National Orchid Garden was mandatory and this is where I first heard the steady hum of cicadas in the treetops, signalling that I was indeed somewhere more tropical than Oxfordshire. The gardens contained some of the most beautiful and perfect orchids that I have ever admired and the diversion of mini waterfalls, ferns and curious palms were enough to keep the visual stimulus occupied.

There was a rather wonderful stone statue of a monkey sitting atop the head of a man, which

Page 15: Orchids, National Orchid Garden, Singapore **Page 16:** Bird of Paradise Plant. National Orchid Garden **Page 17 top:** Lunchtime in Singapore **Page 17 below:** A beautiful temple

Page 18 top: Serious about leisure!
Page 18 bottom: Skewered Tokay Gecko anyone?
Page 19: Beautiful stone tiger mosaics

Page 20 top: Downtown Singapore
Page 20 bottom: Thian Hock Keng Temple

reminded me of the saying I use at work when it comes to not having more than one line manager. I always advocate 'one monkey; one back' and I smiled broadly at the sight of the comical figure before me.

I was very impressed with the indoor collection of bromeliads and the accompanying explanations of what they were, including their distribution, plant habits, pollination and dispersal, adaptations and indeed economic importance.

Other spectacular flora included gigantic bamboos and the bird of paradise plant, which seems to set the hearts of many a botanist aflutter.

I am by no means a plant expert; rather an enthusiastic gardener; but I do so love exotic and tropical plants, which unfortunately are either difficult to procure in England or are not adapted to survive for long in our climate. I couldn't help wondering what it would be like to have these plants around my own home, breaking up the day with colour, design and sheer beauty.

We left the gardens with the intoxicating fragrance of a thousand flowers in our nostrils and the colour of the same plants entrenched in our hearts. The next stop would see us alight at the Thian Hock Keng temple, or the temple of Heavenly Happiness, which is the oldest Chinese temple in Singapore. This Taoist/Buddhist temple was built circa 1820 and dedicated to Ma Zu Po, the mother of heavenly sages and the protector of sailors. Apparently, it was built by seamen grateful for safe passage in the days when the waterfront was evident there.

Lunchtime beckoned and Angie's stomach thought her throat had been cut (she is the queen of the one-liner is Angie) so we traversed the little roads in search of a restorative meal. We passed another beautiful temple and found a little café situated in a dubious looking back street. We sat at a table on the pavement as people strolled by, going to or coming from destinations unknown. The food was pleasant and the waitress managed to eventually understand what we were trying to order. To be fair, we were well off the tourist track so expectations of someone speaking English and understanding our accents were put in perspective.

I speak German and the girls can manage French and a little Italian but Mandarin was a step too far although it really is a language that people should learn as it is a growing currency.

Bellies full, we endured a long and hot walk to find Chinatown. Luckily Sue remembered it from years ago so we knew we had the right area – the markets. The girls purchased rather fetching kimonos and Angie proved to be a dab hand at haggling. She comes from Liverpool so I fully expected her to score points in this area. They don't take prisoners there and they do expect a fair price. Sue is rather more sanguine about haggling but will do so if she really wants something badly enough.

As tomorrow would be a day to visit the zoo and then fly to Australia in the evening, I took my last chance to make a purchase in Singapore. I selected three neck ties. One had the Chinese dragon on it, one had killer whales, and the last had fighting zebras. There are times when an explanation into the machinations of my mind simply won't suffice so I won't bother justifying these purchases.

I am especially proud of my dragon tie, though. Under the Chinese calendar I was born in the year of the dragon. We were informed by a Singaporean lady that mothers always want their daughters to marry people born under this sign.

It is interesting to note that the dragon is the only mythological creature. All the others signs are real and include the monkey, rat, ox, rabbit, horse, tiger, snake, goat, rooster, boar and dog. Joan of Arc was a dragon, as were John Lennon and Mae West. Dragons were believed to land on people's houses (softly, I hope) to keep them safe. They are the most powerful and divine of creatures.

There were a couple of other things worthy of note. The first was humorous. I could not help but laugh aloud as I chanced upon a shop selling meat. Blinking through the rain, which was now coming down quite hard, I saw a large neon sign on which were the words 'Piggy Porky'.

Secondly was the sad sight of two large boxes of tokay geckos which had been killed and skewered on sticks, I can only conclude to be used as medicinal curatives. There was no meat as such on the bodies; merely skin and the intact heads and tails. Having seen these engaging little reptiles alive many times I found it profoundly saddening, but again, who am I to judge?

A very apt name indeed

A Daintree Diary

Tuesday 4th November

Singapore Zoo

I had been looking forward with great anticipation to visiting Singapore Zoo. When it comes to the question of being pro- or anti-zoos I have to say I would rather have them than not. Many rare species are bred and born in zoos and many are saved from the bullet.

That said, I have seen some appalling conditions in zoos, particularly for animals like polar bears and I do not agree with keeping an animal contrary to its natural habitat. I mean, polar bears are just not meant to exist in hot and humid conditions; they are polar animals. Some birds need more room to fly, some fish need more room to swim and yes, there are big questions for many of the animals.

But Singapore Zoo is very good indeed. Set in 28 hectares of land, it is one of the most outstanding zoo settings in the world. It was opened in 1973 and founded by Dr Ong Swee Law whose dream was to build a zoo for Singaporeans to enjoy in peace and tranquility. This would be enhanced by tropical gardens and a rich diversity of animals. People were to get close up with nature and learn to care for it.

It is a place of learning, and there is particular emphasis on fragile forests illustrated through displays of many invertebrates (whose group make up to 95% of all known animal species on earth). It does not take the uneducated visitor long to understand that modern humans are dependant on rainforests for survival and they are encouraged to 'take home some thoughts' on environmental and conservation issues.

One of the signs in the zoo reads:

> *The forest is a peculiar organism*
> *of unlimited kindness and*
> *benevolence, that makes no*
> *demands for sustenance and*
> *extends generously the products*
> *of its activity; it affords*
> *protection to all beings,*
> *offering shade even to the*
> *axeman who destroys it*
>
> Lord Buddha, 500 B.C.

I thought I would entertain you with fascinating facts about some of the animals. A sloth, when it decides to move, will take about a month to travel 1km. Tree kangaroos are just that – arboreal (tree-dwelling) kangaroos that spend 60% of their time sleeping.

There are a multitude of questions for a good natural history quiz night so try these…

A Daintree Diary

- What is a red lory?
- Which parrot hangs upside down when sleeping?
- How long on average can a polar bear swim for?
- Where do you think the endangered golden-headed lion tamarin originates from?
- How many eggs does a green anaconda lay at one time?

Find out later what the answers are.

I was privileged to see the maned wolf *(Chrysocyon brachyurus)* from South America. This beautiful canid has very long legs to allow it to see above the tall grass in its natural environment. It is omnivorous and does like the occasional free-range chicken, much like our own foxes. I arrived at the enclosure at feeding time just as the warden was lobbing large (dead) white rats to the occupants. There were two wolves and they didn't panic and didn't fight. They calmly scoured the long grass for the meal and I was impressed by the stature and poise of this animal. When you ask if zoos are good you have to consider what would happen if these were the only two left in the wild.

I have not forgotten about those questions I posed. The red lory is a bird. The parrot that curiously often hangs upside down when sleeping is a blue-crowned hanging parrot. The polar bear can swim for some 10 hours at 10km/hr and the golden-headed lion tamarin is found only in Brazil. Finally, my trick question – anacondas do not lay eggs; they give birth to live young.

I cannot end without talking about the 'man of the forest'. I refer, of course, to the orangutan. It's the world's second largest ape and also the largest tree dwelling mammal. There are two species: the Bornean *(Pongo pygmaeus)* and the Sumatran *(Pongo abelii)*. It is incredible that few realise that orangutans are seriously endangered as a result of habitat destruction, poaching and forest fires. The Sumatran species is critically endangered now. There are thought to be around 7000 left on the whole planet! This animal is the flagship of the Singapore Zoo and some 31 have been bred successfully and sent all over the world as part of breeding programmes. What else can one do? Unfortunately I don't have the answers but I do know that we have to keep trying and supporting those who are desperately working to save these and other animals. Several orangutans are roaming pretty much 'free' at the zoo and I managed to get close up to a trio who had come down to the keeper for a cool drink. They are beautiful animals; so gentle and so harmless to us. If they disappear, the world will be a dreadfully poorer place for it. Let us accentuate the positive and hope that initiatives worldwide serve to support and increase populations. I recalled an excellent title of a book on endangered animals called *Now you see me, soon you won't*.

Sadly, it was all too soon that I had to leave to make ready for the flight to Darwin that evening. As I walked across a small bridge to the exit I observed a strange creature lurking down below, just visible in the semi darkness of the shade of a tree. It looked like a crocodile but – hold on – it had a very thin snout. I was looking at a false gharial *(Tomistoma schlegelii)*. Little is known about this freshwater reptile and it is already endangered. It is native to Indonesia,

Page 25: A palm, resplendent in the sunshine
Page 26 top: Butterflies feeding
Page 26 bottom: A huge bat – known as a flying fox displays his wares.

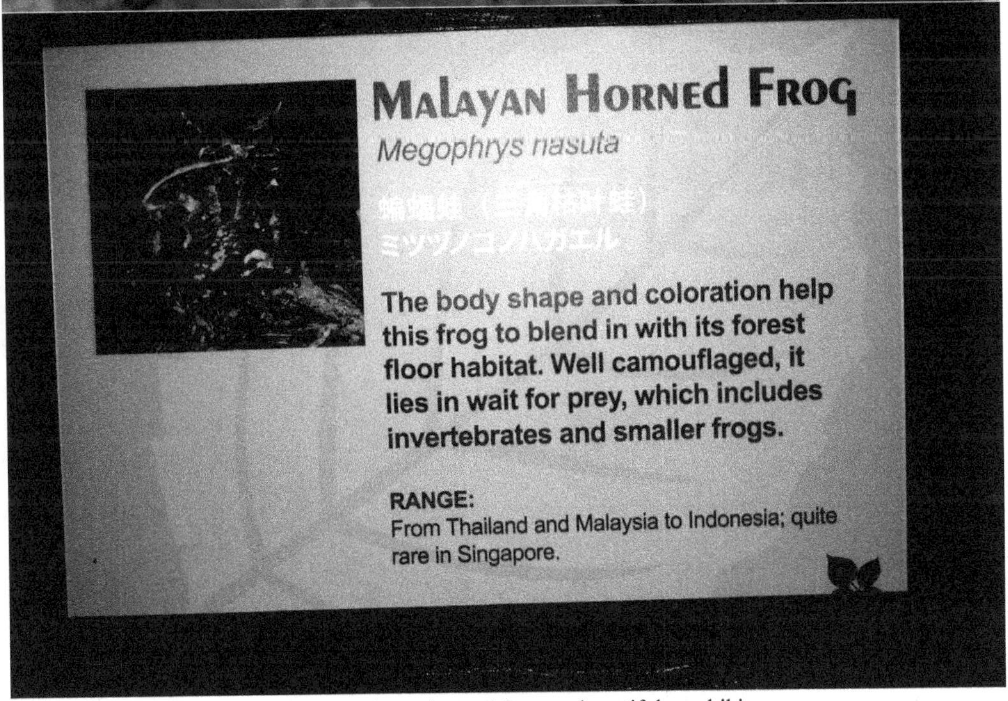

MALAYAN HORNED FROG
Megophrys nasuta

角蟾蛙（三角枯叶蛙）
ミツヅノコノハガエル

The body shape and coloration help this frog to blend in with its forest floor habitat. Well camouflaged, it lies in wait for prey, which includes invertebrates and smaller frogs.

RANGE:
From Thailand and Malaysia to Indonesia; quite rare in Singapore.

Page 27: Malayan horned frog – a beautiful amphibian
Page 28 top: Orang Utans feeding
Page 28 Bottom: A thorn spider feeding
Page 29: Plenty of food to go around

Malaysia and very possibly, Vietnam. It has a sharp mouth and elongated, slender jaws perfect for catching fish. It will take macaques if one is careless enough to pass close by. Distribution data is incomplete but there are thought to be no more than 2500 left so to see a live one was a rare privilege.

It was stiflingly hot and I returned to the hotel to find the girls had been shopping and had followed this with a good swimming session in the pool. I got the beers in immediately (that's what blokes do) and we relaxed for a while before setting off to the airport, Australia bound. The dream was being lived and I closed my eyes to imagine what it would be like to once again be in a rainforest. The last time I did so was in Ecuador in the 1990s and the wait had been far too long.

I can report another good flight to Darwin and then to Cairns later at night. Customs was very tight indeed not just because of the tragedy of 9/11 and the continual threat of terrorism, but the Australians don't take too kindly to people bringing flora or fauna in or indeed taking any out. I fully understand this standpoint. I feel the same about England. We want to keep out the Colorado beetle and offensive people, thank you very much.

We went through not just one but several security scanners. The gentleman at the receiving end looked at his colleague and said "That's the one" pointing to my black bag. I thought 'What do you mean, "that's the one," mate? There's nothing in there. Except there was! The security man opened the bag, looked at me and said "What's this?"

I looked down to see a green tree snake in his hand.

I explained that I liked natural history and I had it as a keepsake. The reader should now exhale gently as the serpent was a self assembly plastic specimen; a harmless toy that I bought from Singapore Zoo. He smiled, looked at me and said "Oh, you like snakes do you? What else do you like?"

I felt like telling him that I liked casual sex with strange men and asking what he was doing later but I refrained. In my usual enthusiastic way I proceeded to tell him that I studied arachnology and once bred endangered tarantulas including the goliath bird-eater *(Theraphosa blondi)*, the biggest species on the planet; the size of a Frisbee.

"Well, you can't take anything out of this country, mate," he spluttered. I explained that I had no intention of taking anything out and that my camera was the only tool I needed for capturing animals during my stay. He calmed down a bit and began to tell me about a huge snake he saw at home. I suddenly realised that the ladies were waiting and I said,

"So am I free to go on my way now?" to which he agreed. I told Sue that I bet they took my name and would give me the full works when I left the country. Rubber gloves and Vaseline sprang to mind and I maintained that hideous image as someone said "Up against the wall!" The police arrived with the sniffer dogs and all the passengers were 'encouraged' to line up with their bags against a wall whilst the hounds did their bit. They seemed almost disappointed not to find anything; the people that is – the dogs couldn't really give a damn.

I know they have a job to do like everyone else at every airport around the world but I maintain that there should be more civility and courtesy shown to passengers. Hey – we are not all criminals. We are real people who fund economies around the globe by visiting their countries. Perhaps it's just me but a little smile or a polite "would you mind?" would go a long way. It's hard being a dreamer some days. Of course when we returned to England there were no checks at all at Heathrow. Anyone could have got through. It was a stark contrast to the other places we went to and we are paying a high price for our liberalism and that's for sure. Why don't we put up a sign saying, 'Welcome to England, just come on through and disappear into the system forever.' My granddad would not have approved. In every other country there was a sense that you were really being checked out before entry. There was a palpable respect for the people doing the job. In UK it's like people say "how dare you ask me to turn my bag out, this is an infringement of my civil rights, my human rights, my something or other rights that I have not found a lawyer to explain yet." That would be enough to get the liberals pleading for them, but in Australia you would have an interview without coffee and be dealt with in no uncertain terms, quite rightly.

Soon there will be complaints about Christmas trees in our own mainly Christian country. Oh, wait – that's already happened. The people that uphold or bow to such ridiculous vitriol should be ashamed. Doubtless they are politicians, councillors or other do-gooders who want the votes and an easy life. My fridge has got more bottle than those idiots.

Chapter 2.
Anyone for cane toad golf?

Wednesday 5th November

It was Bonfire night in Britain but here we slept almost throughout. I drifted into a deep sleep as I arrived at the apartment. My last thoughts were of my first images of Australia as we wound our way from the airport to our dwelling place. There had evidently been a fire through some parts of the forest that lined the twisting roads. I thought this conflagration to be a natural occurrence but I later learned it had been started deliberately and controlled by the state. There were plants there that were far from dead amongst the ashes, such as the remarkable cycads, which we shall come to later. We also caught our first glances at the Great Dividing Range which is the fourth longest range of mountains in the world at some 3,500 kilometres. It runs along the entire eastern coastline and was originally the home of aboriginal tribes.

Our residence (or base camp) was the Mandalay Apartments in Port Douglas. These are a series of self-catering apartments owned by Derek and Raelene Hall, and managed by a small team of marvellous people. Nicole and Kira were also key players who provided an efficient and friendly service. They had all the information a visitor could want – even details of Australasia's premier nudist resort, which was imaginatively called the White Cockatoo. I thought I would give this one a miss but it was always there as a filler should I feel the need to go really wild down under.

Our apartment had two separate double bedrooms, a spacious living room and a good kitchen where there was plenty of space to prepare food. There was a balcony on which we would be having breakfast and other meals and this looked out onto a field, and a small road used to traverse to and from the apartments. There were several tall trees including giant bamboo immediately outside our balcony. Many gentlemen would shout "hooray" to the idea of living in the same apartment as two girls for a month but it has its drawbacks, and for them as well. Luckily, we all knew each other and agreed that should any of us be seen walking about the apartment disrobed, there could be nothing we have not seen before anyway. Besides, we would be out and about for much of the time and frankly, we didn't give a hoot. The moment we opened the balcony doors we were met with a cacophony of screeches from within the trees. Once our eyes had focused we could see the perpetrators – beautifully coloured lorikeets. They were flying around, chattering and squealing amongst the trees. It was a wonderful sight: no cages; no bars. I told Angie a joke – I said a mate of mine had been sacked from a pet shop because he was caught with his hands in the trill, but I knew I would have to do better than that with a Liverpudlian who comes from the land of comic greats such as Ken Dodd.

Also those cicadas were making a right din, which delighted me greatly. And there were a few other birds with beautiful songs but I was unable at that time to ascertain what species they were.

As for Port Douglas, it is some 70 kilometres north of Cairns. For the anoraks amongst you, its co-ordinates are 16° 29 ″01 ′S 145° 27 ″55 ′E. The township was established in 1877 after gold was discovered and it quickly grew to a population of 12,000, with 27 hotels. However, when the nearby Kuranda railway was built the town dwindled and a monstrous cyclone flattened the town in 1911, all apart from two buildings. In 1960 Port Douglas was little more than a fishing village with around 100 people.

Tourism began to boom in the 1980s and it had around a thousand residents in 2006, although this number increases hugely in the summer months. It is situated near to the Barrier Reef and has the rainforest nearby so there is much to tempt the inquisitive traveller. Many Australians take their vacations at Port Douglas.

On July 5th 1943 a RAAF Vultee Vengeance aeroplane (serial number A27-217) crash-landed onto the beach near Port Douglas. In November 1996 U.S. President Bill Clinton and his first lady chose the town to stop on their historic visit to Australia. We saw his photograph at the little restaurant he must have eaten at. There is a lovely little story that whilst eating there he witnessed the wedding certificate of a couple who had got married the same day. Apparently he was there again, eating in the Salsa Bar and Grill on the day of the 9/11 attacks so he had to leave abruptly.

I was deeply saddened (almost traumatised) to hear that one of the world's great conservationists, Steve Irwin, AKA The Crocodile Hunter, died at Port Douglas when he was stung by the barb of a stingray at Batt Reef. It was a million to one chance. The ray felt threatened as it was surrounded by cameramen as Steve swam above it. It must have felt that it had no escape and it lashed its tail, plunging its defensive barb through Steve's chest, piercing his heart and killing him almost instantly. Ironically, he was filming a series called The Ocean's Deadliest at around that time. News of his death shocked the world and there were even 'revenge killings'; mutilated rays were being found on beaches prompting Steve's close friends to ask for the disgusting practice to stop.

Like many people, I knew that Steve Irwin was an entertainer and played to the cameras, but so what? He exposed the beauty of Australia and its animals and plants to the world in a way unique to the man. I remember some of his catchphrases such as "You little beauty" or "Crikey" or "Have a go at this." My favourite, however, was "Ooooooh, he's gettin' cranky." It's a little known fact that he had a fear of parrots after being bitten many times by the birds.

After all those close encounters with crocodiles and cobras, and more besides, who could possibly have thought that he would have left us the way that he did? He was only 44 years old. Wherever you are, Steve, may you be surrounded by the beauty, majesty and glory of the flora and fauna that you loved in life.

Thursday 6th November

Let's get one thing straight: apart from tapeworm and turnips, March flies *(Cydistomyia doddi)* are the most revolting and annoying things on this earth. They may be better known as horseflies and the adult females feed on mammalian blood – especially human. Now I have been out with a few females willing to feed on my blood (after the contents of my wallet have gone) but this winged sentinel takes some beating. The ubiquitous insects took any opportunity to land on my face, neck and exposed legs ready for a tasty bite. I got tagged in the small of the back by one and it really smarted. I was more annoyed that the perpetrator of this heinous crime got away before I had the chance to splatter the contents of its ugly frame across my back. I can console myself with the fact that it will be dead by now and I am still here, eating crumpets and imbibing strong but delicious tea.

There were also sand flies, which are tiny buggers that you don't see until you have been bitten and your skin starts to itch and redden, and a large yellow and red lump appears. Then we have the highly irritating mosquito. There is the dengue mosquito *(Aedes aegypti)*, introduced from Africa. It also feeds on mammalian blood but the males feed on nectar. They transmit dengue fever and dog heartworm. Next up is the common banded mosquito *(Culex annulirostris)*, which bites at night and transmits such wonderful diseases as Ross river fever, Murray Valley encephalitis, and dog heartworm.

Finally, the giant mosquito *(Toxorhynchites speciosus)* is around but thankfully both sexes feed on nectar. The larvae are predators of other mosquito larvae, which is good news indeed.

That's quite enough of the Diptera for today. The girls had a dip though, in the pool. The water was very warm which was unsurprising given that the temperature that day was almost 35 degrees. I didn't bother with the pool for two reasons. Firstly I have the body of a stick insect, and whilst I am not totally paranoid about it I would rather not give the kids in the pool something to laugh at or the resident canine a reason to believe that there was a light lunch for the taking.

I never had muscular arms. I tried when I was younger to 'pump some iron' and eat two steaks a day. All it did was cost me more and leave me knackered. For heaven's sake, I even worked on a bloody farm. Not one ounce did I put on. I recall in the early days of the Internet receiving two separate spam emails. One assured me that I could 'lose 4 stones' in weeks and the other insisted that Viagra would be of great benefit to me adding several inches to my member. Now just imagine what I would look like walking around at seven stones with a permanent erection? I should think I would be arrested for some kind of public disorder. Toppling over would be a real possibility though that might just be wishful thinking on my part.

The second reason was more practical. I figured that since I had flown some 10,000 miles to get here I would at least bathe in the beautiful blue ocean. After all, you can go into a mere pool down at your local swimming baths. However, there were serious restrictions to going into the sea; but more of that anon.

It was a day for energising and preparing for all the experiences to come. The girls got plenty of sunshine, and swimming and reading time whilst I had plans to cram in as much photography and natural history adventures as I could. This was Australia, the land of rich diversity and I wanted to absorb as much as I could in the short time we would be there. I looked like one of the famous five in my shorts and shirt.

The day turned into a beautiful evening and I walked along Four Mile Beach with Sue. There was a warm breeze caressing our faces and cool moist sand beneath our feet. There was hardly anyone on the beach and it felt like our own. It is worth noting that for some obscure reason, Four Mile Beach isn't actually four miles long at all. A local beach walker told me so but I am unable to recall how long it is. If you want to go and measure it feel free to let me know the specific distance.

As we were walking, Sue asked me a rather strange question. She enquired "does the sun still rise in the east and set in the west down here?" Instantly I replied that it did but it set me thinking. It's one of those questions like 'if a tree falls in the forest and there is no-one around, does it still make a noise?' Well – of course it bloody does…doesn't it? I looked into this upon my return home and the boffins tell me that the sun rises in the east and sets in the west everywhere on earth except for the exact north and south poles where there is no east or west. To Australians however, who are south of the Equator the sun rises a bit north of east and sets a little north of west. Got that?

It was time to eat and we elected to visit one of the Chinese restaurants on the main Port Douglas strip. I saw the photo of Bill Clinton from the time of his visit and I could not help but smile. "I did not have sexual relations with that woman," he said. I imagined him not paying the bill and saying "I did not walk away from that table" but I am being unfair. At least Clinton was attractive to women, unlike some of our own dour characterless leaders. Imagine any of our great 'politicians' on the lawn at number ten with….no hold on, let's not even go there. As we waited for the meal to arrive we spotted the little pink geckos zooming around the walls and ceiling. They were lovely creatures with jet black eyes and feet that look and act like suction pads although they are nothing of the sort. They actually have fine filaments which interact with the surface they are on at a molecular level. I wondered if I was fast enough to catch one. For sure, Steve Irwin would have been. The meal itself cost some $94 which we considered to be expensive. If we were to spend that just on an evening meal alone in the next three weeks we were going to spend a lot of dough, so we agreed that this was a treat and we would limit such fiscal frolics to certain evenings. We walked off some of the meal and returned to the Mandalay Apartments very happy after an excellent day. I had not finished, however, as I had been straining at the leash to get out there with head torch and camera to see what creatures of the night could be found lurking underneath logs and in the trees.

Page 35 top: Queensland is a beautiful place **Page 35 bottom:** A red hot ginger plant
Page 36 top: A storm brews over the Marina in Port Douglas **Page 36 bottom:** Plenty of boats for fishing or leisure **Page 37:** There are no Health and Safety rules at the market
Page 38: A strange plant in Port Douglas

It was still very hot outside. I took my trusty Nikon D40 and torches and set off around the grounds. I almost immediately stepped on something distasteful. Not what you might be thinking of – dogs were nowhere to be found, but a rather large and feisty *Bufo marinus* – the cane toad.**

I should explain a little about this much hated amphibian. It was introduced into northern Australia in 1935 (believed to be from Hawaii, although the species originated in South America) in order to kill the cane beetle, which was eating the sugar cane crops. The idea was that the toads would eat the beetles, the crops would be fine and everyone would be happy - except some bright spark had not done their homework. The cane beetle lives primarily at the top of the cane and the cane toad does not climb so one was never exposed to the other. However, the cane toads thrived in the tropical climate and there are now well in excess of 200 million of them.

There have been numerous attempts to control them in the past. Indeed, one of my new-found friends informed me that the government introduced an initiative where people were rewarded for catching cane toads by exchanging large quantities of them for beer. This sounded great to the FNQ's (Far North Queenslanders) and they set about the task night and day with gusto. The problem was that the land was full of pissed up Aussies stumbling around the countryside out of their heads on the amber nectar. This was dangerous to one and all and the spoil-sport government called the whole thing off. I was further informed that around that time cane toad golf was a favourite pastime and all of the golf shops ran out of nine irons. Let it not be said that the Aussie is not innovative and they do love their sport. I also saw many cane toad purses in shops. They looked pretty awful and I could not see one of the old girls down the Labour club bringing this ugly article out of her handbag.

There is a more sinister side to this. The cane toad has a voracious appetite and eats almost anything it can stuff down its throat. It lives for up to fifteen years and it also has poison glands, which are highly toxic to most of the local wildlife. Huge quantities of indigenous species such as the northern quoll *(Dasyurus hallucatus)* have declined in numbers as a result. Indeed, the poison, Bufotenin, is a class one drug under Australian law – the same classification as marijuana and heroin. I asked a couple of zoologist friends in the area what was being done about it. They were none too happy that the government was, in their view, doing nothing and it seemed that if it wasn't in their backyard they didn't want to know. It doesn't look good. I have no idea how they can control this destructive creature. It is with some relief, though, that I can report that nature itself is finding a way. Certain snakes know the toad and leave it alone; rats are beginning to learn that if they turn the toad over, there are no poison glands underneath so they can eat it from the belly. It will be interesting to see what else evolves to combat this pest. Before I visited Queensland I would have laughed at the cane toad issue but now I have seen the beast first hand I genuinely feel for those people. It's a stun-

---

** MANAGING EDITOR'S NOTE: This was where I had a minor insurrection on my hands. The author quite properly wants the cane toad to be known as *Bufo marinus*. However, the technical editor Max Blake wanted it to be known by its more recent synonym *Rhinella marinus* which is seen as correct in the more *avant garde* corners of taxonomy. I just shrugged my shoulders and remembered when it was known as *Chaunus marinus* and ruminated on the fact that zoologists are very strange creatures, and wished I hadn't given up smoking.

ningly rare and beautiful area and something needs to be done. I hope with all my heart that they eradicate *Bufo marinus* before it's too late. There is one shining light at the end of this disturbing tunnel. Reports in the *Daily Telegraph* newspaper in England say that there is a 'tiny solution to Australia's woes'. Apparently the native meat ant can attack and devour baby cane toads. The venomous skin does not affect the ant. I offer this information with a little caveat, however. The article appeared in the April 1st edition 2009 and whilst it seems to be a serious article, I would not rule out some kind of April fool spoof.

In any event this is a stark lesson in reaping what you sow. Disastrous results can occur by introducing non-indigenous species to an area without proper thought of the consequences. I spent the next hour or so wide-eyed and happy in my own little world.

I found several species of spiders, large cockroaches and large brown geckos. Finally, before I retired I stumbled across a rather bewildered preying mantis.

Friday 7th November

I was really excited about today, but that will be no surprise to the reader as I always seem to be in some state of excitement. It was time to get into some real rainforest and visit the Mossman Gorge, which is the largest part of the Daintree National Park, covering some 55,000 hectares. It contains a wealth of plant and animal life and some of the oldest plants on the planet. Both girls were joining me and the group on this trip, and I was a little concerned how they would fare. They had not exactly packed for the rainforest (flip flops are about as useful as boxing gloves in this environment) and indeed, Angela had been having trouble with a foot injury. However, I have to admit to being impressed. She bounded around like a kangaroo on caffeine, resplendent in those natural rainforest colours of green and brown.

Okay, I lied there. She had white trousers with a purple top, a pink hat and shoes and a bright yellow rucksack and was therefore as easy for an animal to spot as a helicopter on your driveway. That said, she and Sue gave it a right good go and I was immensely proud of them. To be fair, Sue had a pink top and bright yellow bag also. What a dynamic duo.

After a short drive northward up the Captain Cook Highway we exited the car and gathered in a little group whilst our guide, Jennifer, gave us a pep talk. Most important of all was the identification of a plant called the leaf stinger. This beauty has leaves that when brushed against even lightly can produce an awesome sting. So much so that I heard tell it stayed with one poor soul for many years and he came out in horrid lumps whenever exposed to water.

The leaves have hairs like slithers of glass that enter the skin and break off, simultaneously injecting some of the irritant. When I tell you that the irritant in question contains formic and acetic acids, histamine and at least one other unidentified chemical you will understand why Jennifer had our full attention. There is no known cure and even the Kuku Yalanji aborigines from the area never fully managed to remove them. Whilst Jennifer was explaining this, I noticed out of the corner of my eye one of the group suddenly jerking his right leg as if performing a rather dubious Highland fling. Then he stopped and jerked again, this time impersonating a footballer kicking a ball, but there was no ball there. This American gentleman was becoming very agitated and I admit to desperately trying not to laugh aloud.

"What the hell is that?" he exclaimed after spinning around like the Tasmanian Devil in the cartoons. It transpired that the unfortunate man was being attacked by a March fly (remember our old friend?) that was so large as to appear to have spent most of its short life on steroids. We briskly moved into the forest to escape the flying fiends.

Our group consisted of Americans, Canadians and Brits. The two Americans were really funny. Openly gay, they proceeded to have little disagreements whilst wearing the most outrageous hats. I mean how could you take them seriously? I am far from homophobic (I really couldn't care less) but they just seemed to be totally misplaced in this environment but credit to them, and once again who am I to judge? In actual fact, I really took to one of them and we spent some time at the back of the group looking for insects that the others had missed. The Canadians were a husband and wife team from Ottawa, a place I had fleetingly visited. I congratulated them on giving us one of the world's finest rock bands in *Rush* and I hoped that they were satisfied with the fact that we had given the world Ozzy Osbourne. They nodded politely but obviously thought I was an incurable freak.

We had to cross a precariously placed wooden suspension bridge at one point – a contraption that Sue positively loathes because of her fear of heights, though she does not fear flying. I did not underestimate how difficult this would be for her as the bridge began to sway and lurch quite radically, but she held on gamely and got to the other side relieved and pleased to have slain another demon.

Jennifer's enthusiasm was contagious and her love of her surroundings very evident. In reminding us that there was less than one per cent of rainforest in Australia she evoked strong feelings of being privileged to be there. She loved her job but the pay was poor so she only led tours occasionally.

We kept hearing a very strange bird and she confirmed it was a rare Victoria's riflebird that was not seen so often. It is a beautiful black bird of paradise named after Queen Victoria. I only just caught sight of it as it flew overhead into the canopy but even that sighting was a lucky experience.

It is about now that I should tell you about the strange creature that is the cassowary. It is a huge flightless bird, a bit like an emu and native only to the tropics in Australia and New Guinea. It is actually the third largest flightless of bird on earth. They have dark plumage with an iridescent blue neck and something called a casque on their heads. This is a mound of hard skin and the purpose of it is not precisely clear, though many have hypothesised about it being a sexual tool, or an implement for bashing through undergrowth. They are shy birds, but capable of inflicting vicious and devastating wounds with their middle toe, which has a razor-sharp claw. They could rip your belly open in a second and I have seen some pretty horrific pictures to qualify that.

They are incredibly beautiful and critically endangered. There are thought to be only about 1500 left in the wild and I remember Jennifer's reply to the question "have you ever seen a cassowary in the wild around here?" She said "I wish." I truly hope she gets to see one but the truth is their habitat is being destroyed by pigs of all creatures. Wild pigs will eat their eggs and much of the food available on the forest floor. The Mossman Gorge used to be their sort of terrain, but sadly no longer.

We sallied forth deeper into the forest taking in the sights and sounds and smells of this truly remarkable environment. I never used to be that interested in botany but a wise man once taught me that in order to get to see the invertebrate life I craved, I would do well to learn what plants they either live in or feed off. This turned out to be a most excellent piece of advice. Now I know which plants tarantulas like to live in, and which plants will best attract butterflies and other insects.

However, I should like to begin with the big stuff. There are a great many species of fig trees in Far North Queensland and they live curious lives. The strangler fig (*Ficus virens*), known locally as 'Jarrangkal', is extremely large and can begin its life in the canopy if a seed germinates on an upper branch. Other fig trees will make their start more conventionally, from ground level. Ripe figs are very important for a variety of animals including parrots, possums and fruit bats.

The hollow roots of some fig trees are used to make boomerangs and these 'buttresses' develop in many shapes and sizes. Trees with buttresses are shallow-rooted and lack a tap root so it is a wonder why in tropical storms or cyclones they don't all just fall over. However, they have as much chance as any tree of remaining upright in such circumstances. As these roots are lateral they are able to take in more nutrients from the forest floor and also avail themselves of oxygen from the air. There are special cells called lenticels, which have been found in mangrove roots enabling the tree to take in air.

The cluster fig has its fruit flowering from the trunk, rather than the branch tips or foliage. This phenomenon is known as 'cauliflory' and it is not known exactly why this occurs. A very plausible hypothesis is that the tree demonstrates close forest living whereby it does not want to present its fruit up in the canopy to such animals as pygmy possums and bats. I am led to believe that this behaviour does not occur in identical trees in subtropical rainforest.

I have come to understand that plants can be exciting. I appreciate their adaptations. As I write this in the 200th anniversary of Charles Darwin's birth I recall the famous phrase so often championed by business gurus… 'It is not the strongest of the species that survive, or the most intelligent; it is the one that is the most adaptable to change.' It is with this in mind that I must write about the marvel of leaf colour. Usually, back home in England or indeed in any temperate forest, dying leaves are easily identified by their red, green, yellow, brown or gold hue. Autumn time displays such resplendent beauty even in a time of dying. However, in the tropical rainforest of North Queensland they don't have seasons. Many plants have a little trick. As I walked through the forest I noticed many trees with pink and red leaves and I assumed they were dying off. Jennifer, wise girl that she is, commented that this was far from the case and

The curious phenomenon that is 'cauliflory'.

The legendary 'wait-a-while'-vine.

the reverse was true. The coloured leaves were in fact the new leaves and the plant was trying to fool caterpillars into thinking they were old leaves and therefore not nutritious. It works on most of the insects apparently. What a brilliant ruse. The rainforest is a vicious battleground. It is a place where if you stand still long enough something will grow on you, live in you or eat you. When one stands to look around, nothing seems to be happening but there are seedlings everywhere poised, ready to shoot up for the light the moment a large tree falls and the sunbeams stream like golden lasers through the canopy onto the forest floor. The seedlings germinate almost immediately then grow to a few inches tall and wait for that one defining moment which may take decades. They have to germinate quickly and at least establish their own spot or they would simply rot in the ground.

I have already alluded to the pain and suffering brought about by an encounter with the stinging plant but there is another formidable opponent waiting for the careless and unwary traveller to experience its sweet caress. The plant known as 'wait-a-while' or lawyer plant is covered in needle-sharp, thick prickly stems and it can ensnare a passer-by easily and you have to

The plants have devious tricks.

'wait-a-while' whilst you unhook yourself from its barbs. It is actually a palm that grows ruthlessly with backward-facing barbs. It probably feels like having an encounter with that medieval torture instrument, the Iron Maiden. Finally, I should be neglecting my duty if I did not highlight the wonder of fungi; in particular, the colourful jewel of the forest, the bracket fungus. These rather bizarre structures are actually an important facet of the litter cycle and their only purpose is to disperse and spore. I observed some orange fungi that was circular in shape

Page 46: Mossman Gorge. **Page 47 top:** Tangled roots of fig trees. **Page 47 bottom:** Boomerangs are made from the hollow case. **Page 48:** A paradise for entomologists

with frilled white edges. I must say it was a rather startling plant and one would have expected to see more but there was not 'mushroom' on the forest floor. I had to get that one in!

Enough of plants and fungi for now: what about the animals? There wasn't much to see as it was daytime and all the sexy stuff happens at night. Snakes wend their way through the forest looking for a meal, rats scurry around, spiders emerge to demonstrate the often productive trait of a sit and wait strategy, and many other animals spark into life as birds warily sleep on the very tips of branches so they have a chance of escaping when a hungry snake approaches and bends the branch, giving the bird a warning. I did see a Boyd's forest dragon *(Hypsilurus boydii)* fixed upright on a thin sapling, motionless and wide-eyed. These beauties are not dragons of course but lizards and quite abundant but only in this part of Australia. They certainly look as if they have come straight from the dinosaur age with their scaly bodies, fleshy folds on their necks, white spines and long, slender tail. They are not very large creatures, perhaps a foot in length, and as with any animal, seeing them in their natural habitat is a hugely rewarding way to make their acquaintance.

I then happened across one of my favourite invertebrates. This insect more than any other sings in my head when I am back at home in England. It reminds me of faraway lands and many enjoyable adventures. It never fails to let me know when I am in its world: a world where noise is the key to survival; not just looks. I alluded to it previously. It is the delightful cicada. One might be shocked how few people have even heard of a cicada, let alone know what one looks like. They are the bugs that create the chirping, buzzing noise from the treetops. They can be found in many lands across the world but alas, not in my back garden during the heat of a summer afternoon in Oxfordshire. How joyful it would be to rid ourselves of the noisy human brats jumping incessantly up and down on trampolines up and down the gardens of England and replace them with such a splendid insect. Sigh!

Let me tell you a bit about them. Firstly, the most amazing fact about at least one species is that it lives in the ground for up to seventeen (yes, seventeen) years before emerging for only a few days to procreate. I believe the Australian species live underground for six to eight years but that's still one heck of a long time. The nymphs feed on the sap of roots. We visited Australia at a time when an army of cicadas had emerged and the empty, moulted skins of their former bodies betrayed the fact that new life was overhead in the trees. Cicadas, like many other animals, shed their skin in order to grow. Many people scoff when I declare my interest in invertebrates but they make up 95% of the animal kingdom and are much more fascinating than the woolly cuddly creatures that have been virtually 'done to death.' But how do they generate such noise? Insects normally rub their legs together to generate sound but the cicada has an organ called the tympanum on the side of its body. It flexes this inwards then it suddenly pops out to make the sound. It is done so rapidly that to our ears it sounds like one regular tone. The sound can be quite deafening when the treetops are full of sex-charged males all trying to outdo their opponents. It's a kind of talent show in the sky. They are eaten in many parts of the world and written into folklore especially in China when the phrase 'shedding the golden cicada skin' means to use deception to escape danger. The adventure was nearly over but there was still the opportunity to swim in one of the freshwater areas of the river running through the gorge. The ladies stripped off (steady lads) and plunged in whilst I opted to stay on *terra firma* and take a few photographs to record their memorable occasion. I have a proclivity for accidents around water and recall falling into a river in Ecuador whilst laden with expensive camera equipment. I also recall ripping open my left foot whilst snorkelling in Malta. I concluded that since there was so much of the trip to go I would be better off staying fit and well to enjoy it all. As we emerged from the forest back to the vehicle a rather tired looking brush turkey hopped into our path. The poor little fellow had a bad leg but the locals all knew him and he was happy to scavenge what he could off strangers. He was in all other respects a splendid bird: jet black with a red head and yellow neck. The tails of these birds are unusual in that they are fan-like in appearance and used to send signals to other birds.

Mossman Gorge is one of those places to see before you die (or before it dwindles away) and I heartily recommend a visit. A large Aboriginal community still resides there and if you are lucky you can join them on an educational tour of the forest. Living there would be a dream. There is no frost – ever – and the sunshine is plentiful. There is no chlorine or fluoride in the water as it is drawn freshly and the local community worked together (remember that feeling, anyone?) to ensure the necessary mechanisms were put in place to serve the people.

Page 50. One can swim in the clear waters of Mossman Gorge
Page 51. A new beginning for these spiders

Page 52 top: The author knows the way to Mandalay
Page 52 bottom: The perfect base
Page 53 top: This *Tetragnatha* spider has huge jaws
Page 53 bottom: Unidentified spider (male in background)
Page 54 top: A beautiful rainbow lorikeet feeding
Page 54 bottom: These geckos were common around the Mandalay
Page 55 top: This large wolf spider hunts at night
Page 55 bottom: After rain frogs can drink from bromeliads

Page 56: Abdominal patterns were very attractive
Page 57 top: There was plenty of room to drive around town
Page 57 Bottom: This flame tree was a beauty

Page 58 top: A green ant carries a wing away
Page 58 below: A mudskipper - peculiar fish which don't seem to realise that fish are meant to live in water

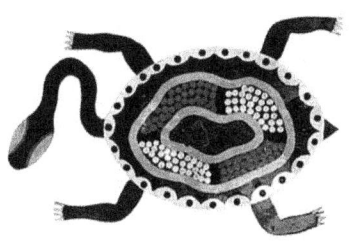

Chapter 3.
Python patrol

Saturday 8th November

We awoke to a blinding sunlight and the sound of lorikeets squabbling in the treetops. One of the trees had fruited and there was an almighty free-for-all going on. I managed to take a couple of acceptable photographs and we enjoyed a fine breakfast of cereal, toast and orange juice on the balcony whilst discussing the plans for the day.

Sue and Angie were off to the reef for a snorkelling session. Since the reef was full of deadly jellyfish they would be wearing 'stinger suits' and I was intrigued to know what they consisted of. I was to be surprised upon viewing the photographs upon the ladies' return but more of that later.

I had other plans so the blue of the ocean had to be sacrificed at the altar of the green forests. Whilst back in England I had made contact with Anne and Mark Palfreyman who manage a farm in a beautiful area of the Upper Daintree. Anne is a genuine and friendly lady who has five-star experience fronting up some of Australia's top restaurants and Mark is a gregarious zoologist who loves his cricket. When not attending to the steers they run rainforest tours but Mark agreed for me to have a private trip into the forest at night, which was just what I had been waiting for.

I hired a car for the drive north and vividly recall the freedom of that journey up the highway with the sun shining, shades on and Jimi Hendrix playing *Angel* on the rather dated radio. I was surrounded by rainforest and ready to find some spiders and hopefully at least one snake. I was particularly keen to locate a python, which were pretty lively at that time of year. It didn't get much better than that trip. It was one of those 'why am I not doing more of this?' afternoons. Thoughts turned fleetingly to Blighty and the ice-cold rain, and I felt good! My thoughts were shattered as I heard a thud on the side of the car as I was driving along. I was shocked to see that a brightly coloured bird had flown into it. I remember thinking it would be damned typical if a rare species had met its end with Carl Portman but I could do absolutely nothing about it. All that space and the stupid bird flew into me.

After a few wrong turns I found a sign for 'Tranquility on the Daintree' and accessed the dirt-track towards my destination. Apparently traversing dirt-tracks in small hire-cars is frowned upon by the leasing company but they were really laid back about it. The young man in the reception said "Don't worry mate, this is FNQ – chill out."

The track went on for miles and the car bobbed and weaved and strained as it left a constant dust cloud in its wake. The landscape was becoming ever more beautiful with an alchemy of forest, creeks and hills. It is worth explaining a little more about this area of planet earth.

The Daintree was added to the World Heritage List in 1988. There are four criteria used to determine whether an area should be included on the list and the Daintree Rainforest is exceptional in that it is one of only twelve sites worldwide that meet all four.

The criteria are:

- It must be an outstanding example of the major stages in the earth's evolutionary history
- It must be an outstanding example of significant ongoing geological and biological processes and man's interaction with his natural environment.
- It must be an example of superlative natural phenomena, and
- It must contain the most important and significant natural habitats where threatened species of animals or plants of outstanding universal value live.

And let us not forget that the Daintree area is adjacent to another World Heritage site of equal importance, The Great Barrier Reef, making it the only place in the world where two natural World Heritage Listed sites meet.

I finally arrived at Mark and Anne's house. It was private land so I would not be meeting any noisy tourists, which was a great relief. It was cool in the air-conditioned car but I emerged into a blast-furnace heat. In a few moments Anne, who was clearly pregnant, appeared with her trusty dogs but one was secured in the garden as he was not fond of strangers; especially tall, thin specimens from Birmingham. However, Rocky was very friendly and often came to me for a petting session.

"Hi, you must be Carl. You're a bit early as Mark is out at the moment but he will be back in a while. You are free to go and explore the forest, take a dip in the clear Niau Falls and generally make yourself at home. Oh, and if you would like to join us for dinner tonight you are most welcome." She also gave me a bottle of water, an ice cream and some dried apricots to take on my walk as she knew I would require sustenance. Now, who says the Aussies are not hospitable? What a lovely greeting.

And suddenly there I was, alone and free in the oldest tropical rainforest on earth. It was exhilarating and nerve-wracking too. If I should get bitten by a death adder whilst on my own it would be serious but it was unlikely that this would happen. They usually live on lower ground, especially by the palm-lined shores of the coast, which was worth remembering. However, I was always careful when poking around under logs, or in grass.

I spent the afternoon observing freshwater crayfish, cicadas, butterflies, lizards, birds, spiders and varieties of plants. I was wowed by the first orb spider I saw *(Nephila* sp.) and the way that I managed to find life in places that one might normally consider to be too barren to support it. The March flies were here in depressing numbers and I was being driven crazy trying to hold steady photographing subjects as the flies bit into me. One hammered me in the small of my back and it really hurt. I lost the lens cap to my camera somewhere and retraced my steps to find it, which I did (against my better judgement) but at some cost. I had drunk all my water and eaten my food and my strength was waning fast so I knew I had to head back.

During the walk back I observed a magnificent Ulysses swallowtail butterfly. Imagine an iridescent blue butterfly with black-framed wings the size of your hand. They are a protected species in Australia and rightly so. They fly high in the rainforest and if you are lucky you will suddenly catch the blue flash of wings zig-zagging through the trees. They live fascinating lives and the female mates only once in her life laying her eggs on a tree called *Melicope elleryana* because this is the only plant that the caterpillars feed on. She lays 150-200 eggs, one at a time. They then live for up to a month as their wings wear out due to erratic flying. They are virtually impossible to see when at rest as the underside of their wings is brown and they are perfectly camouflaged. I arrived back at the car to be met by a well built bearded gentleman with a smile as wide as the Daintree River. "Hi mate," he said and I responded in kind.

There be crocodiles here

"You must be Mark," I ventured to declare, and he confirmed that he was. We chatted about several issues ranging from events in England to what we were going to do that evening. I was very grateful to him and to Anne for giving up their time so that I could have fun in the forest.

Mark confirmed that we would venture out at dusk (which was not too far away) and hopefully find what I was looking for: at least one python and large huntsman spiders, which don't appear in the daytime. They are crepuscular creatures. I told him that he could go about his work and leave me to explore as I did not wish to take too much of his time. As dusk fell I visited some of the trees in the very large garden and it was not long before I saw a large brown huntsman spider resting on the bark of one of them. I have always fought an uphill battle with spiders as so many people despise them. The fact is they are much more afraid of us than we are of them. I had to approach with caution as any sudden movement would see the spider dart back behind the bark. Carefully I moved in with my camera and took my first photo of a large Australian huntsman. They don't seem to mind flash at all so I snapped away to my heart's content.

"Hey Carl," shouted Mark from his house. "Look what I have found".

I rushed over and he pointed to the guttering where a most outstanding white-lipped tree frog *(Litoria infrafrenata)* was resting. I recalled again how life would be if such wildlife could be found around my own home. Mark never took it for granted, however. His spacious bungalow home is built in prime forest and the doors are always open as it never gets below fourteen degrees there even in the middle of winter. As a result he and Anne get lots of visitors and I

A stunning white lipped tree frog *Litoria infrafrenata*

observed a variety of frogs and geckos on the floors and walls mercifully keeping the mosquito population down. They also had reptilian visitors in the form of snakes and he had removed a brown snake from the eaves of his roof only a few days previously. I am not too sure I would want those around my house so perhaps I should be happy with the squirrels back home.

The plan was to have dinner and then venture out but an important job had to be done first. We jumped into his large white pick-up truck and had a drive around the area to see if there were any snakes on the road – but had no luck. We stopped at one point and looked over a valley through which ran the river and his cattle grazed by its banks about a mile or so in the distance.

"Listen mate; I need to move those cows because they are too near to the water's edge and they will for sure be attacked by crocodiles that populate this area." He said we would go back to base and he would ride his quad bike whilst I drove the pick-up truck and the plan was for him to go down to the river and drive the beasts up to higher ground where I would be with a cordoned-off area to see that they went into the safety of a field for the night. His pick-up truck was old and battered but I soon found out how valuable it was. The terrain was very tough with marshy land, hillocks, pot-holes and streams, and the vehicle traversed it all with ease. I thoroughly enjoyed playing my part and earning my corn as part payment for Mark and Anne's hospitality for the privilege of experiencing the forest.

As for the crocs, Mark informed me that there were plenty of them out there and indeed a neighbour had in recent days lost his dog down by the river's edge. Half of the canid was found on the bank, which is a very sad thing but a reminder that this is a part of the world where walking out at night is a very dangerous activity.

Mark's enthusiasm for wildlife is such that he takes any opportunity to indulge. Before dinner he grabbed a couple of pairs of binoculars and we headed off to a nearby field to see what birds we could find. A fig tree had fruited and many species had been visiting it that day. Most had gone when we arrived but we were lucky enough to catch site of some channel-billed cuckoos *(Scythrops novaehollandiae)* or fig hawks, as they are also known. These are beautiful birds that look nothing like the cuckoos we see in Europe. They grow up to 67cm and are pale grey in appearance. They possess a very thick, straw-coloured, downward curving bill and red skin around the eyes. Imagine a big grey crow, crossed with a parrot and you have the idea. These impressive birds parasitise the nests of Australian magpies and crows.

There were white-bellied sea eagles in the area too. These are the second largest bird in Australia after the wedge-tailed eagle. They are fairly common above coastal rainforests and a truly magnificent sight on the wing.

"Argh; shit!" exclaimed Mark, suddenly. I thought he had seen something dramatic.
"What's up; what is it?"

His answer made me laugh aloud. "We've just lost another bloody wicket."

Australia were playing India at cricket. In fact they lost to them and he was none too happy. There was always some cricket commentary going on in the truck or on his TV at home. This reinforced the notion that the Aussies take the game very seriously. We had a brief conversation about cricket but I don't profess to knowing that much about it. We both asserted that our teams would win the ashes in 2009 (we *did*) and he reminded me that England cannot win in India either. I was simply out of my depth trying to find a winning argument but had it been football I would have had him! I told him I supported the mighty Aston Villa and he declared that he had actually heard of them but only because they played 'his' team, Manchester United. We agreed that he could occupy the higher moral ground on his chosen sport and I would do so on mine. It was a wise decision and we maintained good international relations.

We hastened back for dinner and Mark made a delicious Spaghetti Bolognese with plenty of garlic and chilli. He and Anne were lucky enough to be able to pick chillies off their very own chilli tree. I was most grateful for this meal, which was made more enjoyable by the excellent company at the table.

At last it was pitch black and time to begin exploring. Mark was unsurprisingly very knowledgeable about the local area and knew where to go to find certain animals. He decided that we should start with a real adventure and go down to the river's edge to look for crocodiles.

Now, seeing crocs in zoos or wildlife parks is one thing as you know there is a barrier between you and them. However, standing on a riverbank in the dead of night with wild crocodiles in the vicinity is a different thing altogether. Mark told me in no uncertain terms to follow him and not stray, and to keep my eyes and ears open at all times. He had a huge spotlight and I supplemented it for short-distance viewing by my own head- and hand-torches.

We drove to an area where he had seen a huge croc some days before and we exited the vehicle just a few metres from the riverbank. It was utterly exhilarating and I tried not to think what would happen if I was grabbed by a hungry reptile. He shone his light across the river looking for the giveaway sign of a croc in the water: red eyes. The crocs' eyes shine red when the light of a torch is reflected back. There were a couple shining back but these were fairly small reflections and certainly did not belong to the bigger crocs. I thought 'If they ain't there, where the bloody hell are they?' as I shone my torch all around. I had never been so acutely aware in any wildlife situation that I was in the animal's territory and not mine.

Suddenly something touched my foot and I jumped a mile. I looked down to see a large cane toad gawping back at me. My heart was thumping within my rib cage and I disliked the toads even more now. Mark gave them no mercy – but then I never found anyone who did. If they were on the road he would drive over them. If they were under foot, they were trodden on. I could not bring myself to hurt one, partially because I was on foreign soil and did not want to interfere but also because I don't believe in taking the life of an animal simply for the sake of it. Nature would have to take its course.

Mark told me more about crocs in the area. A local man was killed when bathing after having a few drinks, and that's how most croc deaths occur. Alcohol generates bravado and this can

Page 65 top: The first huntsman I saw on the Daintree
Page 65 bottom: Tranquility on the Daintree – typical terrain

Page 66: A freshly emerged cicada resting
Page 67: The glorious Australian huntsman

About Huntsman Spiders

These splendid creatures are often commonly known as 'tarantulas' or giant crab spiders. They are not tarantulas but 'true' spiders with sideways working fangs rather than forward pointing. They are very flat and this design enables them to enter cracks and crevices. A very large specimen could reach 160mm but the males are usually smaller than the females. Some are considered dangerous to humans but the vast majority are completely harmless. They are nocturnal creatures with keen eyesight and they catch prey by springing on it or literally chasing it down. These spiders can be found in family groups and the mother is especially protective of the young. They are found throughout Australia in forests and in deserts. Not all Aussies will agree but I think they are outstanding fauna.

Page 68 top: One of the many species of frogs
Page 68 bottom: A lizard clings to a vine at night
Page 69 top: This spider was hunting even with several appendages missing
Page 69 bottom: A huge white kneed king cricket *Penalva flavocalceata*

Page 70 top: Bracket fungus
Page 70 bottom: 'My' all terrain vehicle
Page 71 top: Spiders will happily live in the garage
Page 71 bottom: The dreaded March fly *Cydistomyia doddi*
Page 72: Mark Palfreyman

cost your life out here. The chap had driven to the river on a hot night, and taken a dip. Another local man drove by, stopped the car and told him to get out of the water as there were crocs about. The drunk decided to do so and was observed approaching his vehicle, which had a door open and the radio on. The passer-by drove off believing he had done his duty and the drunk was safe. The next day police found the car with the door still open and the radio still on – the man was never seen again.

Further up the river a man had been killed after returning to lobster pots that he had put out. That was not his mistake – his error was returning to the pots at the same time each day. The crocodiles cottoned on to this and were waiting for him. He was ripped apart in minutes. They aren't stupid.

Finally, Mark had been looking after a group of German visitors whose four-wheel drive broke down in the middle of the river whilst trying to cross it. Now, I know what I would do; I would wait in the vehicle until help arrived. Unfortunately the men decided to get out and try to push the jeep through the river, splashing about as they did so. Mark happened to come by and gave the men a fearful rollicking and told them they should consider themselves very lucky to be alive.

As I write this on February 18th 2009 I have just read the following story on the Internet:

> 'Boy, five, killed by crocodile
>
> Police have confirmed that a boy who vanished from an Australian river edge was taken by a crocodile.
>
> Remains of five-year-old Jeremy Doble were found in the stomach of a 14-foot male crocodile trapped in the flooded Daintree River- an international attraction for ecotourists - near where the boy had vanished on February 8 (my birthday-Carl) a police statement said. Jeremy had been playing with his seven-year-old brother Ryan and their dog behind their family property in a flooded mangrove swamp when he disappeared.
>
> Ryan told police that he saw a crocodile immediately after but did not see an attack, police said. A 10-foot female crocodile was trapped last week but later released after a surgical procedure found no evidence that it had been responsible for the attack.
>
> Police were unable to say what will happen to the male crocodile, whose stomach contents were similarly examined by a non-lethal surgical procedure.

Crocodiles have been protected by federal law since 1971 and their numbers in Australia's tropical waters have steadily grown.

However, authorities are allowed to destroy crocodiles that threaten humans.

I always follow a golden rule – to be aware of your environment. Respect what lives there and behave accordingly. This isn't a stroll down your shopping high street, this is a potentially life threatening hobby at times jaunting through rainforests and visiting dangerous animals at night.

We never got to see the large croc; it commanded a large area of up to 50kms so was probably further upstream, but we did see a couple of others in the water and I had the memory to take away with me. I will never forget standing at the water's edge with every sensation buzzing, listening to the chirruping crickets and looking up at the night sky, which was emblazoned with bright twinkling stars undiminished by any man-made light. I recall thinking that this was what I was alive for and if I died on that spot I would have gone a happy man. Oh yes – and the crocs would have had a meal too!

The author getting to grips with an amethystine python

Morelia amethistina on the alert

We drove back home, parked up and set off on foot to the rainforest. We had only ventured about 100 metres when Mark suddenly exclaimed "Python!" and there sitting in front of us was the most beautiful, freshly moulted Amethystine python *(Morelia amethistina)* that you ever saw. At this time of year Mark saw about one a week and we were lucky (again) to have come across one. Pythons are not venomous in any way but they can and do bite readily. They kill their prey by crushing it. With each exhalation the victim takes, the snake squeezes a little tighter until death becomes a blessed release.

This chap, however, seemed pretty calm and did not attempt to bite. We took a few photographs, admired it for a few minutes and then set it free to go on its merry way. I handled it from the tail end, always keeping its main body on the ground so it felt safe, much as Steve Irwin used to do in his fine documentaries. He certainly educated millions of people like me the world over.

You can own a big car, you can have a large house and you can have lots of money but for me, nothing can compare to seeing such magnificent creatures in the wild. I recall the saying, money can buy you the dog, but it cannot buy you the wag of its tail.
The next couple of hours saw us finding a myriad of invertebrate life including the white-kneed king cricket *(Penalva flavocalceata)*, which can grow up to 80mm in length and scavenges around on the forest floor.

I was also delighted to find the largest huntsman spider on the trip and photographically recorded it for future reference. Interestingly, spiders seemed to be the only creature that Mark was not too keen on so it was good of him to keep foraging around for them on my behalf.

A cicada heads for the treetops

There were no scorpions to be found, which was the only disappointment but there would be more opportunities later on the trip to try to find some. I knew that several species lived in Far North Queensland, such as the little brown scorpions *Liocheles* spp., but finding any was proving to be difficult.

Again I found myself listening to cicadas and staring deep into the night sky. It is times such as these that I begin to contemplate life, and if there is a God. I recalled what Charles Darwin once said – "*I cannot persuade myself that a beneficent and omnipotent God would have designedly created parasitic wasps with the express intention of their feeding within the living bodies of Caterpillars*". He did have a point there.

All too soon I found myself saying goodbye to Anne and Mark and driving back down the highway towards Port Douglas. I was drenched in sweat but had a camera containing some excellent pictures.

When I got back into the apartments the girls were abed but of course I woke Sue up to regale her with the details of my heroic adventures wrestling with snakes and stalking crocodiles and huge spiders the size of your hand. Well, come on; if fishermen can tell the odd fib about the size of their catch, I can embellish my stories just a little.

Before slipping into a deep sleep, Sue told me that their day snorkelling had been great fun and I was shown a photograph of her and Angie in their stinger suits.

I am unable to show the photograph in this book in case the fashion police pay me a visit or I am reported to the care in the community authorities. They looked like a couple of inflatable dolls with realistic moving eyes. It was a sight I can never forget and I hope that the nightmares will stop sometime soon. I am still paying for therapy to help me cope.

Sunday 9th November
I suspect that both Angela and I have had better starts to the day. To say we had a disagreement would be the euphemism of the holiday. I have a particular sense of humour and Angie, like any lady, is sensitive to dubious comments about clothes. What I said as a joke she took

as an insult and it spiralled from there. Unfortunately poor Sue was the one getting caught in the crossfire. Clearly Angie and I thought it best to be out of each other's hair for the day and she left the apartment for some time on her own. I was furious but assured Sue that she would not be the one in the middle and that I would make a monumental effort to be more civilised in order to keep the peace. I am sure Angie did the same.

We got Angie on the mobile and we all agreed that a half day at the Crocodile Farm between Port Douglas and Cairns would be the perfect opportunity to get us back on track. Instead of us getting snappy with each other we would let the crocs do it!

I had hired another vehicle and we were soon on our merry way. I should point out that when driving in Australia the indigenous population can be more thoughtful than us on the road. Now some of you may find that hard to believe but it is true. They have special 'pull over areas' on twisting roads to enable other drivers to pass. I didn't cotton on to this until I had disparaged several drivers behind me giving it plenty of horn, which made me drive even slower. Hey – that's what happens in Britain!

Hartley's crocodile adventures came into being in 1980 and after many years of hard graft and habitat conservation it has become an award-winning site. There is a crocodile lagoon, a cassowary walk, a wildlife discovery trail, a crocodile farm and the latest attraction, the Gondwana Gateway, which is best described as a bridge between two worlds.

We took the boat cruise and watched crocodiles lunge out of the water to take food that the captain offered on long poles. Here we saw examples of the famous 'death rolls' and headshakes of the crocodiles, which are in-built strategies to finish off their prey. They had saltwater and freshwater crocodiles. By far the most dangerous are saltwater crocs, which patrol the mouths of rivers and coasts. They are the largest of the living reptiles and rely on stealth alone to feed. I was shocked to hear that in winter when the temperature is lower and the crocs eat less, they can survive easily on the equivalent of only one chicken a week. When we see these truly beautiful animals taking prey on our TV screens it is easy to assume that it is an almost daily occurrence when in fact it may take days or weeks to obtain a meal. They are therefore opportunistic feeders, a bit like spiders and snakes.

Angie has a son called Paul so she was very amused to find a 4.6-metre croc with the same name, with the following message on the sign at the enclosure.

'Cattlemen whose properties bordered onto the Annan river near Cooktown had run out of patience with Paul by 1989. He was responsible for taking a number of dogs and had an annoying tendency to bite the heads off cattle when they were drinking at the river.'

He was between 50 and 60 years old and weighed 700kgs. There were all sorts of crocs there; young and old. Some were so old they were covered in green slime and had barely any teeth. They also had various lumps bitten out of them from previous encounters with other crocs and they were now spending their time simply awaiting the next meal.

As we cruised around, two sea eagles flew overhead and I was impressed by their size and grace. Once back on terra firma we went to see the cassowary and other animals in the park. I was really keen to visit the Gondwana Gateway, which had monitors, snakes, eastern water dragons and quolls amongst other specimens. Even crocs have predators (at least of their young) and I enjoyed the opportunity to get up close and personal with a Merten's water monitor *(Varanus mertensi)*, which feeds on crocodile eggs and young. This is another indigenous animal that has been greatly reduced by cane toads in recent years. I should explain as best as I can what Gondwanaland was, or at least as I understand it, since it is very important to understand where the continent of Australia came from.

Gondwanaland was a supercontinent joining East and West Gondwana that existed millions of years ago. It separated from Laurasia nearly 200 million years ago and as Laurasia went north, Gondwana moved south. It included most of the landmasses in today's southern hemisphere, which included Antarctica, Africa, South America, Madagascar, New Zealand, New Guinea and Australia. The name is derived from Sanskrit and means 'forest of gond.' It began to break up even more over time and Antarctica, India, Madagascar and Australia began to separate from Africa and South America, which drifted west. More break-ups ensued over several million years resulting in Australia separating from Antarctica some 80 million years ago in the late Cretaceous period. So now we have animals that were on the supercontinent of Gondwanaland being broken up and moving with the continents. Apes went to Africa and other places, whereas there are none in Australia. Such isolation for Australia shaped the flora and fauna that it has today and it truly has a remarkable diversity. So much of Australia is uninhabitable for humans yet life exists, as it has for millions of years. Evolution is extraordinary at the best of times, but to see animals such as the cassowary makes you stop and consider.

The event that most impressed me about the day was Susan handling a snake, which she says she would never have dreamed of doing before she met me. I was honoured and I admired her bravery and willingness to give it a go. The same would be said later when she was photographed handling a young crocodile. We lunched at Lilies restaurant overlooking the lagoon and I tucked into burger and chips, which was not good for my health but who cares, sometimes you need to pig out and not worry about it. I don't suppose walking near wild crocs was that good for my health either. We had a fine old time after our difficult start to the day. Angie got some really excellent shots of the crocs and demonstrated her dexterity with the camera. Sue was a bit quiet today, probably troubled by earlier events but handling the snake bought her back to life with a smile – which is when she truly shines. She often refers to herself as a 'mad yampee cow from the Black Country' and I can't really argue with that. For those not residing in the British Isles, 'Yampee' means mad or crazy and the Black Country is in the Heart of England in the West Midlands. It was called the Black Country not because it is a country full of blacks (as is commonly thought) but that the buildings used to be blackened with smoke and grime during and after the Industrial Age in Britain.

Page 79: Susan the snake charmer **Page 80 top:** The cassowary is flightless but potentially dangerous
Page 80 bottom: Facing extinction? **Page 81 top:** Lovely lashes
Page 81 bottom: These toes can easily dismember you

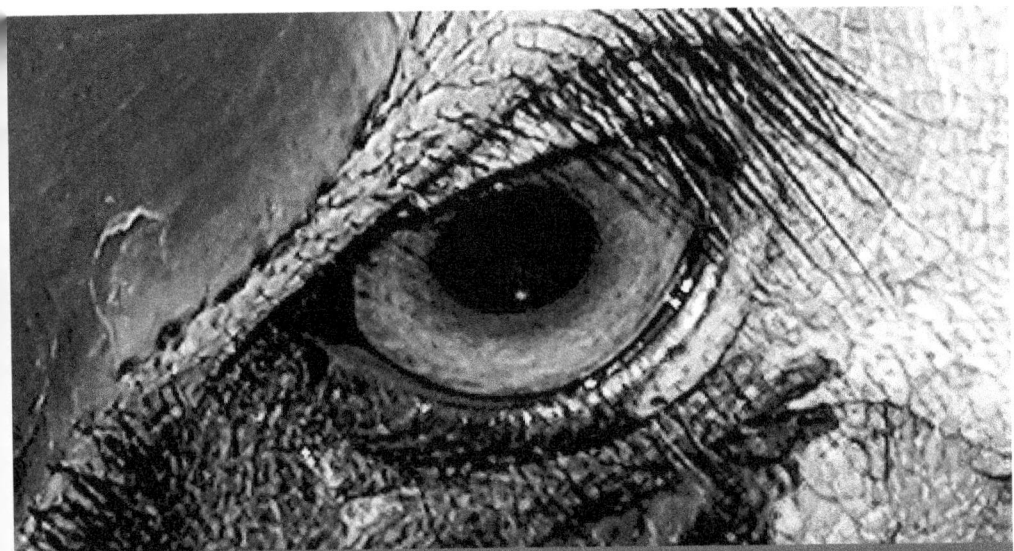

Cassowaries (from the Malay name *kesuari*) are part of the ratite group, which also includes the emu, rheas, ostrichs, and kiwis, and the extinct moas and elephant birds. There are three extant species recognised today and one extinct:

- *Casuarius casuarius*, southern cassowary or double-wattled cassowary, found in southern New Guinea, northeastern Australia, and the Aru Islands, mainly in lowlands.
- *Casuarius bennetti*, dwarf cassowary or Bennett's cassowary, found in New Guinea, New Britain, and on Yapen, mainly in highlands.
- *Casuarius unappendiculatus*, Northern cassowary or single wattled cassowary, found in the northern and western New Guinea, and Yapen, mainly in lowlands.
- *Casuarius lydekki*, Extinct

Page 82 top: A crocodile lurks in the shallows **Page 82 below:** Lizards love to bask in the sunshine
Page 83 top: A curious reptile looks back at the camera
Page 83 below: Carl and Susan with their 'baby'.

Page 84: The Papuan frogmouth *Podargus papuensis*
Page 85 top: A rare gharial, the long snout is adapted to eat fish.
Page 85 bottom: A small heron hunts for fish
Page 86: The flora in Queensland is striking

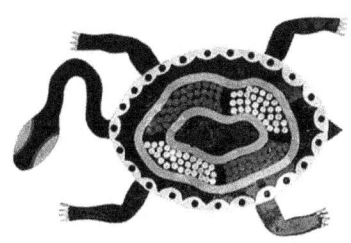

Chapter 4.
Spiders really do eat birds

Monday 10th November

We awoke to the mellifluous sounds of a little songbird, which reminded us that we were due to have 'lunch with the lorikeets' today whilst Angie intended to take in the sunshine and chill either in the pool or at the beach. We agreed upon a half-day trip to the Rainforest Habitat Wildlife Sanctuary, which was nearby and there was a specific reason why I for one wanted to go, which I shall come too presently.

For the traveller who does not have time to see a range of habitats and animals in the wild, this is exactly the sort of place to visit. There are wetland, grassland and rainforest areas packed with plants and animals to see. There are also education stands where you can learn about certain animals and have your photograph taken with them. Sue and I elected to have a photo with a rather scruffy and bewildered looking koala and a small crocodile. It is a great wildlife experience and a chance to observe, really close up, the range of flora and fauna in Australia. Their strap line is '*Through Observation, appreciation, through appreciation, Conservation.*'

We arrived at lunchtime (always a good strategy) and had already paid for a hot and cold lunch buffet where we were able to interact with water birds, curlews, cockatoos and of course those lovely lorikeets, which are mischievous in the extreme but very beautiful.

The food on view was excellent. There was a choice of salads and potatoes with reef fish, eggs, bacon, different breads, and exotic fruits and pastries. I heard a member of staff curse loudly as a cattle egret made off with a huge steak that it had stolen from an unattended plate. It's not good to feed the birds, especially with this, as it is not exactly a natural food source. Still, we cannot blame the bird – we must blame the idiot who allowed it to happen and left his plate unattended.

A gentleman we later knew as Peter - one of the senior members of staff - walked by with a couple of lorikeets and they flew onto our shoulders. Mine began to nibble at my ear, then at my upper lip. Rainbow lorikeets have blue heads, green backs and tails, red and orange breasts and blue bellies, topped off with red bills. They look like they were caught in the blast of an explosion at a paint factory but they are tremendously eye-catching and you can't see the birds when they are in the trees, which seems very surprising to most people. You hear them long before you see them.

Two blue winged kookaburra's gaze at each other

There were many other birds to appreciate that day. My favourite Australian bird is the blue-winged kookaburra *(Dace leachier)*, which is also known as a barking jackass or howling jackass. To me these are stunning birds, streamlined and colourful with a spine-tingling call that sounds a bit like a barking dog. There are laughing kookaburras too but these take the biscuit. Another bird that must be seen to be believed is the obscure Papuan frogmouth *(Pod Argus papooses)*, which is nigh on impossible to spot when at rest. I recommend that everyone reading this should Google this bird and marvel at its construction.

There was another bird I wanted to find as I had seen one outside our apartments flying from tree to tree. It was olive green with a yellow breast, white belly, black tail and a black bill, with yellow under the throat. I looked around and could not find it until that kind gentleman Peter took me to where it would be. I looked in the trees and there it was: the figbird *(Sphecotheres viridis)*, which does indeed live in open parks and gardens as well as on rainforest edges. It's a beauty for sure and I hoped to see the one near our apartments again. Before

Flying foxes resting in the trees

we left we had the opportunity, at last to see what I had been waiting for. Every night at dusk a huge colony of flying fox fruit bats fly for several miles in search of food and they reside by day in the trees at this rainforest park.

It was fairly breezy and they could be seen swaying from the highest branches and chattering noisily with fights breaking out here and there. There are several species of flying fox: the black one, the spectacled one and the little red one. These were one of the first two; probably the spectacled flying fox. These truly astonishing fruit and nectar eaters have a one-metre wingspan (imagine that!) and fly for up to 25-35kms per hour over many kilometres to their favourite trees. They are vital for seed dispersal and pollination. I can understand why many of the locals want rid of them from their patch but I don't condone such a move in any way shape or form. The bats were here first – let them stay. For some reason females can fall victim to a paralysis tick, which almost immobilises them and they fall to the ground. The Far North Queensland Wildlife Rescue team is always on hand to help.

We drove back home having had a thoroughly enjoyable day and decided to hit the swimming pool. Angie had indeed succeeded in having a lazy day chilling and had a lot of beach time. She joined us in the pool for a refreshing dip and all was well with the world. After supper I ventured out again to stumble across the largest cane toad I had seen to date. We looked at each other for a minute, both trying to suss out what the other's next move was – then we parted company and disappeared into the night.

Tuesday 11th November

Since we had a big day looming on the twelfth we decided to dedicate the whole day to restoring body and mind and spent time in the sea, in the pool and just sprawled here and there, reading. I was writing copious notes up and still trying to find the figbird, which I never did see again. I did pop to town to the post office to send some mail. It was there that I asked a rather asinine question. I enquired as to how much the post would be first and second class to which the lady replied "There ain't no first and second class here; there is one class and it's out of the back door into the van." That told me. Bloody Pommies eh?

The day was quiet and reflective, however, something exciting was happening and a story was breaking in the media about a giant spider eating a bird. Way out in the Atherton Tablelands, a retired gentleman by the name of Les Martin captured some astonishing photographs of a spider eating a bird. A lady in a local shop told me about it when I was talking about spiders – she believed it was a tarantula or whistling spider as they call it in many parts of Oz. Mr. Martin told the Cairns post that it was "an awful thing, the spider was just chewing into its head. The spider's head was going up and down, and it was gouging into him (the chestnut-breasted manikin) at the top of his beak. It was still wrapping it up and then the spider just left it. It was like it was too big or something". The photos were sent around the world via email and I have reproduced one in this book.

It was indeed a shocking incident for the world, but not for me or anyone who has a deep interest in arachnology. Firstly, I need to clarify a couple of points. The spider was not a tarantula but one from the genus *Nephila* or an orb web spider. These spiders can build huge webs, which are very strong and can easily catch birds, bats, small mammals and amphibians. Spider silk is stronger than steel, pound for pound, and Cana-

Spiders will eat whatever they can overpower – in this case, a bird.

dian police once had a bullet-proof vest weaved with spider silk, which withstood a bullet and did the job. The spider would have left the bird wrapped up to avoid injury. It will have injected venom and simply waited for it to take effect. Why get drawn into close quarter combat and risk injury when you can simply strike, walk away and come back when the prey is dead? Many venomous snakes behave in a similar way.

Well, it was a great story and I wondered if I would be fortunate enough to come across a unique wildlife moment in my remaining time; perhaps a human arm sticking out of a hungry croc's mouth, a taipan striking out at its prey or even a kangaroo punching out the lights of a local drunk. Unless you are out there with your camera these things just pass you by. I was ready for anything.

Susan was not ready for anything, however. She was not all at ready for the horrific sunburn she suffered today from less than an hour sitting on the beach. She had put sun cream on, but after going into the sea for a dip forgot to cover her back and she was wearing a swimsuit. When she took it off back at base there was only a small white outline of skin that the garment had protected, against a background of scarlet flesh. She cursed her lack of awareness and knew that when the blisters came it was going to hurt like hell. She was not wrong on that account.

Carl and Sue in buoyant mood.

Angie, clearly at home on the beach

Chapter 5
Above the canopy

Wednesday 12th November

Kuranda is quite literally a village in the rainforest. It lies some twenty-five kilometres north-west of Cairns and there were many reasons for the three of us to visit today. Quite apart from the fact that Angie wanted to meet some real aborigines (it's Australia, after all) and Sue wanted to sample the atmosphere of the place and take a trip on the Kuranda scenic railway – I had an appointment with the Australian Venom Zoo. Then there was the incredibly absorbing 'Skyrail' experience where one can enjoy a 7.5km cable-car ride over the top of the rainforest, gaining an 'above-the-canopy' view. Sue and Angie didn't really fancy this but since I had flown from the UK they thought it only fair that they should make sacrifices too. In truth it is very easy to be paralysed by fear in a small cable car but they overcame any psychological hurdles and I applaud them for doing so.

Now for a little history: this part of the tropical north rainforest is home to the Djabugay aboriginal people who have lived here for over 10,000 years. Exploration by Europeans in the early 1800s opened the way for gold prospectors and the timber industry and settlement by the pioneers. Kuranda was first surveyed in 1888 by Thomas Behan, and the building of the railway and the road from the new seaport of Cairns paved the way for trade and the movement of people over the mountains.

At a suitable elevation of 380 metres above sea level coffee was the crop of choice until severe frosts in the early 1900s wiped out the harvest. Kuranda became a destination for locals on holiday and honeymooners, and word soon spread telling of the magnificent Barron Falls and the lushness of the rainforest. During the 1940s there was a significant military presence in the area; training and rest and recreation for troops and Air Force personnel took precedence over tourism.

In the late 1960s Kuranda was the place to be, having as it did, spectacular scenery, a wonderful climate and cheap living so you could grow your own food and do your own thing. So-called 'hippy communes' flourished for a few years. In the 1970s new settlers arrived: musicians and people with artistic talents and imagination pursuing an alternative lifestyle. Their unusual hand-built houses of brick and timber were inspired by this unique place. Open-air market stalls sold locally grown produce and an abundance of handmade wares. Buskers and fortune-tellers entertained the crowds. The community prospered. The population grew fast with the improved road allowing commuters to work in Cairns and live in the clean atmosphere of Kuranda.

Kuranda today has a variety of restaurants and cafés. It is a very laid-back place but stylish and sophisticated in its own way. I certainly felt able to relax at once. The shops are full of handcrafted wares and aboriginal artefacts, and you really do feel as if you are taking a step back in time when you are there. There are nature-based tourist attractions, which include koalas, kangaroos, reptiles, butterflies and exotic birds. Then there are the Barron Falls and you can even take an 'army duck ride' through the forest. I have to admit to not knowing what this actually entails, but I never had the time to engage in it either. Answers on a postcard please.

Now to the wonder that is Skyrail. The Skyrail Rainforest Cableway experience spans 7.5kms over Australia's pristine tropical rainforests. It is possible to glide just three metres above the verdant rainforest canopy before descending deep into the heart of the forest. It's a thrilling experience. The original Skyrail concept was conceived in 1987 and was followed by seven

Page 94 bottom: A carriage car glides above the treetops
Page 95 top: A view of the canopy, seldom seen.
Page 95 bottom: Climbing ever higher above the forest.

Page 96 top: The tree bark is stunning against the foliage
Page 96 bottom: Trees flower and spread their leaves towards the light

Page 98 top: Time for a photo stop at Barron Falls.
Page 98 bottom: River and rainforest meet.
Page 99 top: The plants came in all shapes and sizes.
Page 99 bottom: This large tarantula is a real beauty.

Page 100: Our comfortable carriage was at the rear of the train, thus enabling us to get the very best view of it as we journeyed through the countryside. Such a civilised way to travel!
Page 101: Steve Irwin would have said *"Crikey. Take a look at **this!**"*

A glorious palm at Kuranda railway station

Obscure plant life on a palm tree.

years of feasibility studies before construction commenced in June 1994.

Thoughtfully, Skyrail's tower sites were selected to coincide with existing canopy gaps, and were surveyed to ensure no rare, threatened or endangered species would be affected by construction. Before construction commenced on the tower sites, the leaf litter and topsoil was collected and stockpiled for reintroduction when construction was complete. Plant seedlings were catalogued at each site, then removed and propagated during construction, and re-planted in their original locations, with the saved topsoil and leaf litter. The towers were constructed in ten-metre by ten-metre clearings, and were placed as far apart as mechanically possible.

The tower footings were built largely by hand, up to five metres deep in some cases, using picks and shovels. There were no roads built during Skyrail's construction. Workers had to walk into the remote tower sites each day, carrying their equipment, which took up to an hour each way. Helicopters were used extensively to assist construction. They carried equipment, materials and cement to tower sites and rainforest stations. The helicopters carried their loads on 100-metre long lines to avoid wind turbulence affecting the sensitive rainforest canopy.

Due to the size of the tower sites, they were difficult to locate from the air, and GPS satellite navigation and radio communication were used to enable the helicopter crews to pinpoint the ground crews and tower sites. Specialist heavy-lifting Russian Kamov helicopters were used to carry towers to tower sites. The towers were flown in, in sections and assembled on site. Some tower sections weighed up to five tonnes each. The cableway haul rope was then laid and tensioned across towers by the helicopters.

The gondolas stop off at various key locations: the Caravonica terminal; the Red Peak Station, which will take you deep in the forest; the Barron Falls Station, where on approach you can enjoy spectacular views of the Barron Gorge, a deep chasm lined with dense rainforest vegetation; and finally on to Kuranda itself. The Red Peak and Barron Falls Stations were designed to blend in with the surrounding rainforest surroundings, minimising environmental impact and were built in pre-existing clearings. Helicopters transported an incredible 900 tonnes of steel, cement and other building materials into the Red Peak Station site. Fifteen months and $35 million later, Skyrail opened to the public on 31st Aug 1995. The cableway was originally installed with 47 gondolas, giving it a carrying capacity of 300 people per hour; however, a $2.5million upgrade completed in May 1997 increased the total number of gondolas to 114 and the carrying capacity to 700 people per hour. What a fantastic feat!

Stretching to 7.5km, Skyrail was the world's longest gondola cableway at the time of completion. More importantly, Skyrail provided people with a unique opportunity and a world first: to see and experience the rainforest in a safe and environmentally-friendly way. I think it took great vision, hard work and a desire to do something that is a cause for good, and I really do take my hat off to them. The Skyrail project and construction required world-first construction techniques and even today Skyrail Rainforest Cableway can say it is the most environmentally sensitive cableway project in the world.

History lesson over; we next went on the minibus to Kuranda. The sun was shining again and the driver, quite contrary to previous individuals, was outwardly friendly and very knowledgeable. He gave us a running commentary throughout the journey, which was illuminating and very helpful. The subject matter ranged from why cycads were so special, to how sugar cane was grown and harvested, the art of controlled forest fires, termite mounds, grasses and the Kuranda Railway itself. The cycads are just the most awe-inspiring of plants. They are survivors of millions of years and are the first plant to sprout a leaf after a major fire. If there has been rain after the fire, however, the grasses are first but in the dry tinderbox-conditions of this tough environment the cycads are masters of the terrain. And hands up those who knew that one planting of sugar cane can provide up to ten crops. The termite mounds were located in strategic areas with regard to temperature, sunlight hours etc. I think termites are fascinating creatures and I intend one day to give serious time to study their life cycle and behaviour.

Presently, we arrived at the foot of the Skyrail, paid for our tickets and waited excitedly (well, I did anyway) for a gondola to arrive. Sue was dreading the experience and I tried to offer some comforting words to little avail. It was our turn and the three of us got in. The Gondola swayed a little and we shuffled about to get the right sort of balance. I was moving here and there to get the best shots I could and Sue was as white as a ghost. Angie was not overly impressed either but managed to keep good humour. It is really strange that although I detest heights the cable car never bothered me one bit. I think it was because I was so focused on doing something else I did not have time to worry about other things.

To be above the canopy is to see a world that many will never experience. Much of the action in a rainforest occurs well above our heads and it was fascinating to observe what kind of foliage, fruits and flowers were on show. It was hot – very hot – but I continued clicking away, making the most of every second; I might never do this again. There were many different species of trees including the ubiquitous figs and eucalyptus. Some trees displayed huge white flowers that opened towards the sun and the stark contrast of thin silvery bark against the bright green foliage was quite frankly stunning.

We stopped off at all the stations including the Barron Falls. If you had a death wish, here was the perfect place to indulge in a bit of diving.

All too soon the ride was over and we arrived in Kuranda with some hours to spare before taking the scenic railway trip back to Cairns, from where we were driven to Port Douglas by our encyclopaedic friend.

The ladies went off to browse in the shops and meet some local aborigines whilst I strode away to the Australian Venom Zoo – an establishment owned by one Stuart Douglas.

When I arrived at the building two men were outside and one (the chap holding a snake) was on the receiving end of a verbal dressing down. The taller of the two men was being clear about reputation and service to customers and also the issue of him having to cover for staff when he was supposed to be having a day off.

"I see this is the perfect time to interject," I said with a smile. "Can anyone take me to Stuart

A Daintree Diary

The irrepressible Stuart Douglas

Douglas?"

The tall chap glowered. "I am his brother; who wants him?"

I explained who I was and that we had exchanged emails inviting me to call in when I came to Kuranda.

"That's all right then, mate. I am Stuart really – nice to meet you." He left with a parting scowl to the bemused member of staff and took me inside.

He had a couple of things to take care of and invited me to have a walk around and take any pictures I wanted. I was standing in a sort of large corrugated building; dark except for the light off the vivariums and other displays. I walked down a steep metal ramp and came across a sacred shrine.

I was faced with a wall full of media literature, newspaper cuttings and photos of one Steve Irwin. Painted onto the grey wall in large letters was 'LONG LIVE THE KING, THE KING IS DEAD – STEVE 2006.' I felt deeply moved being reminded again of the loss of such a man and recalled how many lives Steve Irwin had influenced and how desperately untimely his death was. Stuart adored Steve Irwin and some of the crocodile hunter's eccentricity and zest for having new experiences was a driving force in Stuart's *modus operandi*.

Tribute to Steve Irwin

There were headlines from the Australian press saying 'Steve falls prey to nature' 'Crikey' and 'Stingray Barb kills Steve Irwin.' One area that I believe the Aussies and the Brits are alike is that even in death, we are able to find humour. There was a newspaper cutting of a

A Daintree Diary

Zanetti cartoon entitled 'Steve Irwin 1962-2006', which showed a crocodile in a pool of water by the river bank, drying its eyes with a handkerchief next to a caption saying 'Crocodile Tears.'

I visited each vivarium observing the contents, appreciating the range and condition of the specimens on display. This part of the world has many deadly animals and I took a close look at the inland taipan, which has the world's most toxic venom. These fellows can grow up to 1.7 metres and it certainly looked impressive. I moved on to the northern death adder, which is prevalent in FNQ and rarely misses if it decides to strike out at a human.

Stuart arrived and greeted me with a cheery smile. He appeared calmer and offered to show me a leaf-tailed gecko, an animal that was much sought-after and often smuggled illegally out of Australia. I held it gently in my hand and it looked back at me through stunning brown and green eyes with a vertical cat-like golden slit.

A stunning leaf gecko.

Moving on, we came to an enclosure with a really sinister-looking snake inside. It was brown with a hint of olive and had obviously just eaten judging by its distended belly. It was the coastal taipan and I was very keen to photograph it.

Stuart carefully lifted the glass lid and I slowly pointed the camera only about four feet away from it. As I was steadying the camera, looking into the blackened eyes of the reptile, Stuart said "Be careful, mate; one bite from that bugger and you will be in a coma within ten minutes."

A recently fed inland taipan
Oxyuranus microlepidotus

If ever there was a time when I needed a lens with vibration reduction it was then. To be honest, when I stop to look at snakes from a different, more intellectual perspective I can appreciate how beautiful and 'fit for purpose' they really are.

I have kept Russian rat snakes and corn snakes but never really taken to them like I do the spiders. These had my utmost respect and undivided attention for obvious reasons. It was time for the last act: a look at the tarantulas with particular emphasis on any *Selenocosmia* species on show.

I got some great photos and saw a range of spiders that whetted my appetite for trying to find that legendary 'barking spider' in the wild later on in the trip. I shall discuss the tarantulas of Australia and the features of this spider later on. I also managed to take some nice photos of the indigenous scorpions that Stuart had. It would have been super to have seen a Sydney funnel web *(Atrax robustus)* but the only specimen he owned had died only the day before. I was gutted and so was he. It was unlikely that I would find one in Sydney.

On my way back to the reception I passed a board full of photographs of Stuart and various animals in the wild. Kuranda has its own wildlife warrior in Stuart. He seems afraid of nothing and no one. He seemed to me to be as happy wrestling some giant snake or snapping turtle as he was holding a scorpion on the palm of his hand. He informed me that his next venture would be in Africa and this well travelled and passionate man will doubtless have a superb time there. He kindly gave me a couple of freebies from his shop and I proudly wear the Australian Venom Zoo cap back here in England. It seems fitting to end this piece with words direct from his website.

> 'At the Australian Venom Zoo our primary focus is on captive breeding and displaying venomous animals that will one day be responsible for producing some of the most advanced medicines in the world. The majority of animals we have here are unique to Australia and in some cases have never before been seen. We believe that the education of the general public is an essential step in understanding the importance of Australia's venomous animals. As well as milking our animals for a sustainable supply of venom for bio-pharmaceutical screening. Our staff are dedicated to the preservation and conservation of Australia's unique venomous wildlife and have appeared in a number of internationally acclaimed wildlife documentaries. Currently, our main focus is on Australian tarantulas, centipedes and scorpions however we will eventually expand to include venomous Australian snakes, jellyfish, cone-snails, insects and other dangerous aquatic creatures. We are the only zoo of its kind in the world and are the only registered breeders of Australian tarantulas. We'd love for our passion to become yours.'

Back into the dazzling outdoors I found Angela and Sue sitting on a step having a quick five-minute rest. I asked Angie if she had found the Abbos. Her reply was less than enthusiastic saying she had come across one solitary man in a shop painting boomerangs, closely 'guarded' by a European lady (Dutch we think) who seemed intent only on trying to get visitors to part with their money on his behalf.

She was bitterly disappointed and Sue was not as impressed with the village as she thought she would be from all the hype. I was happy because I had fulfilled my own mission but felt for the girls. I decided to go and visit the man himself and on the way I passed a dead rhinoceros beetle, which is several inches long. I reached to pick it up and jumped as Angie snapped out "Don't touch that; it's got germs."

We looked at each other and I said "I study entomology Angie; it's my hobby" and she realised what she had said. We all burst out laughing as Angie explained it was her mothering instinct that made her say it. I put the deceased beetle in my pocket to photograph later that evening back at base.

Jim 'Boongar' Edwards

And so it was that I met Jim 'Boongar' Edwards. I was immediately taken with his beaming smile and twinkling eyes and even more impressed with his snow-white hair and beard.

Jim comes from the Wakka Wakka tribe of South East Queensland and makes Didjeridus amongst other things. He is a talented artist and has had exhibitions in Brisbane, Amsterdam, Nijmegen and Cologne.

I could not afford to buy one of his beautiful artefacts but I did purchase his little book all about didjeridus.

I used to think they were spelt 'didgeridoos' but I have used the spelling that he himself has used, which is good enough for me.

I wonder how many words one can derive from didjeridus in anagram form.

I digress - let me tell you a few things about didjeridus then; and I knew almost none of this.

- They are made from eucalyptus trees.
- They are between four and six feet long
- They are hollowed out by termites

- A mouthpiece of beeswax from the native stingless black bee is fitted for comfort.
- A didjeridu that has been hollowed out naturally has got ant channels inside (you can feel them) and these give the instrument its unique resonance.
- Stories abound about where the instrument originated. There is a nice piece of folklore from a place called Mornington Island, where the instrument is called the 'larwah.': The rainbow serpent had refused to let his little sister (a bird) shelter. She got angry and grabbed a torch and burned his humpie (sic) down. The sound of the larwah is compared to the sound that the rainbow serpent made when he rolled around in pain.

Despite the annoying Dutch lady, it was a privilege to meet Jim and I wondered what stories he could tell (given the chance!) of his childhood and the changes he had seen.

After some refreshing drinks we made our way to the railway station to get the train back. This was no ordinary train ride, mind you. For all you railway buffs out there, this is the bit you have been waiting for.

The Kuranda Scenic Railway is world famous. By visiting it you will discover the pioneering history of the tropical north from way back in the late 1800s, be astounded by a magnificent engineering feat and explore some of the great characters involved in the construction of this great railway. The setting was the prolonged North Queensland wet season of 1882. Desperate tin miners on the Wild River near Herberton were unable to obtain supplies and were on the verge of famine. The boggy road leading inland from Port Douglas was proving impossible. As a result, the settlers at Herberton raised loud and angry voices and began agitation for a railway to the coast.

Coming general elections and increasing cold weather in the south saw visits to the north by leading politicians all promising a railway. In March of 1882 the Minister for Works and Mines, Mr. Macrossan, announced the search for a route from the Atherton Tablelands to the coast. He commissioned Christie Palmerston, an expert bushman and a most colourful pioneering character, to find a suitable route.

In February 1882 both Port Douglas and Cairns formed Railway Leagues and engaged in a long and bitter fight for the right to the railway. Not long after, Geraldton, later named Innisfail, entered the competition boasting the sound virtues of Mourilyan Harbour.

During that year Palmerston marked several possible routes from the coast, inland along the Mossman River, the Barron Valley from Cairns and the Mulgrave Valley. In November 1882 Palmerston made the trip from Mourilyan to Herberton in nine days and repeatedly came across the track that had been marked by an inspector named Douglas in May of that year. On arrival Inspector Douglas had wired the Colonial Secretary: 'Arrived Mourilyan 28th May. Fearful trip. No chance of road. Twenty days without rations, living principally on roots. Nineteen days rain without intermission.' Eventually, Cairns won the railway bid.

In March 1884 a surveyor named Monk submitted reports from investigations carried out on all the routes marked by Christie Palmerston. This culminated in a decision that would shape the future of North Queensland. The Barron Valley gorge route was chosen. The storm of indignation that followed from Port Douglas and Geraldton was as enormous as the jubilant celebrations from the people in Cairns.

Construction of the Cairns-Kuranda Railway was and still is an engineering feat of tremendous magnitude. This enthralling chapter in the history of North Queensland stands as testimony to the splendid ambitions, fortitude and suffering of the hundreds of men engaged in its construction. It also stands as a monument to the many men who lost their lives on this mammoth project. On May 10th 1886 the then Premier of Queensland, Sir Samuel Griffith, used a silver spade to turn the first sod. Celebrations involving almost the entire population of Cairns lasted all that day and long into the night, apparently. Construction was agreed by three separate contracts for lengths of 13.2km, 24.5 km, and 37.4km. The line was to total 75.1km and surmount the vast Atherton tablelands leading to Mareeba. Sections one and three were relatively easy to locate and construct. But the ascent of section two was extremely arduous and dangerous due to steep grades, dense jungle and aboriginals defending their territory.

The climb began near Redlynch, 5.5m above sea level and continued to the summit at Myola with an altitude of 327.1m. In all, this section included fifteen tunnels, 93 curves and dozens of difficult bridges mounted many metres above ravines and waterfalls.

Section one of the line ran from Cairns to just beyond Redlynch. The contract was won by Mr. P.C. Smith for $40,000. However, work was dogged by bad luck and a possible lack of supervision. Sickness and disease was prevalent amongst the navvies and the working conditions in the swamps and jungles were approaching unbearable. In November 1886 P.C. Smith relinquished his contract for section one. It was taken over by McBride and Co. but they too had packed it in by January 1887. Section one was finally completed by the Queensland government.

On January 21st 1887 John Robb's tender of $580,188 was accepted for section two. He and his men tackled the jungle and mountains not with bulldozers, jackhammers and other modern equipment; but with strategy, fortitude, hand tools, dynamite, buckets and bare hands. Great escarpments were removed from the mountains above the line and every loose rock and overhanging tree had to be removed by hand. It was during this type of work that the first fatal accident occurred. At Beard's Cutting a man named Gavin Hamilton stood on the wrong side of a log as it was being rolled into a fire, and was killed. Earthworks proved particularly difficult. The deep cuttings and extensive embankments that were removed totalled a volume of just over 2.3 million cubic metres of earthworks. The Barron Valley earth was especially treacherous. Slopes averaged 45 degrees and the entire surface was covered with a 4.6 m-7.60m layer of disjointed rock, rotting vegetation, mould and soil.

During construction navvies' camps mushroomed at every tunnel and cutting. Even comparatively narrow ledges supported stores – some even catering for the men's need for groceries and clothes! Small townships were thriving at number three tunnel, Stoney Creek, Glacier

Rock, Camp Oven Creek and Rainbow Creek. Kamerunga, at the foot of the range, boasted no fewer than five hotels. At one stage 1500 men, mainly Irish and Italian, were involved in the project. Faced with poor working conditions, on April 20th 1888 a meeting of predominantly Irish workers at Kamerunga resulted in the formation of the Victorian Labour League. Even so, relationships between workers and contractors remained harmonious as all realised the magnitude of the task before them. In August 1890 the great maritime strike spread to the railway workers and they formed The United Sons of Toil. They made a demand for 90c. per day. By September differences had been resolved and the navvies' wages were increased from 80c. per day to 85c. per day. By April 1890 Stoney Creek Bridge was almost complete and the project was paid a vice-regal visit by the Governor of Queensland, General Sir Henry Wiley Norman. To His Excellency's astonishment, John Robb prepared a full banquet atop Stoney Creek Bridge with tables, food and wine, dizzily suspended many metres over the gorge. History records that there were no speeches that day due to the roar from the waterfalls.

By May 13th 1891 rail was laid to the end of the second section at Myola. On June 15th 1891 Mr. Johnstone, one of three Railway Commissioners at that time, opened the line for goods traffic only. Just ten days later the Cairns-Kuranda Railway line was opened to passenger travel. Trade at Port Douglas died off rapidly and the town became a quiet little retreat. However, today it is a popular holiday destination. Geraldton (Innisfail) prospered in its own right because of the growing sugar industry. With a reliable supply of goods and freight the Tablelands bloomed into a wealth of rich grazing land. And Cairns was destined to become the modern international tourist centre it is today, still expanding in leaps and bounds.

An historic journey along the Cairns-Kuranda Railway.
MARCH 1882
Search for a suitable route announced by the Minister for Works and Mines
MARCH 1884
Barron Valley route chosen
SEPTEMBER 1885
Cabinet approved working plans
MAY 1886
Premier of Queensland begins construction
NOVEMBER 1886
P. C. Smith relinquished contract for Section One
JANUARY 1887
McBride and Co. relinquished contract for Section One
JANUARY 1887
John Robb's tender for Section Two approved
APRIL 1888
Victorian Labour League formed
1888-89
The first large scale reclamation of swamps at Cairns
APRIL 1890
Inspection by the governor of Queensland

AUGUST 1890
The United Sons of Toil formed
APRIL 1891
The first ballast train reached Kuranda
MAY 1891
Rail was laid to the end of Section Two
JUNE 1891
The Cairns-Kuranda Railway line was opened
1915
Opening of Kuranda Station as it stands today

When we boarded the pretty train we sat in the very last cabin at the rear. The windows were open and we anticipated a cooling breeze as we advanced through the countryside. I will never forget the wait for the train to pull away. The girls and I were first in the cabin and we watched it gradually fill up with various characters whilst listening to the most dreadful Australian acoustic songs I have ever had the misfortune to subject my ears to. The singer was probably a nice guy but this was clearly a form of aural torture. It was a sort of cross between *Always Look On the Bright Side of Life* and *Old Shep*. It wasn't pretty, folks. As we pulled away some kind soul felt pity and turned him off.

We traversed rainforest, rocks, ravines and streams. The high point for everyone on board the fourteen or so carriages was when we arched around in an almost complete circle so if you were in the end of the train, the whole of the train could be seen before you, arching around the fields. What a joy it was then to be the furthest back and therefore able to get a photograph of the whole train in one shot.

I think the ladies had fun but that sunburn of Susan's was breaking out into yellow blisters right across her upper back and some of the lower back too. One had to wince just to look at it and my job was to ensure that soothing lotion was applied. Various 'remedies' were suggested but as most of us know, nothing works – you just have to ride it out. Writing this diary in February 2009 the marks are still clearly there.

The dead rhinoceros beetle that I found.

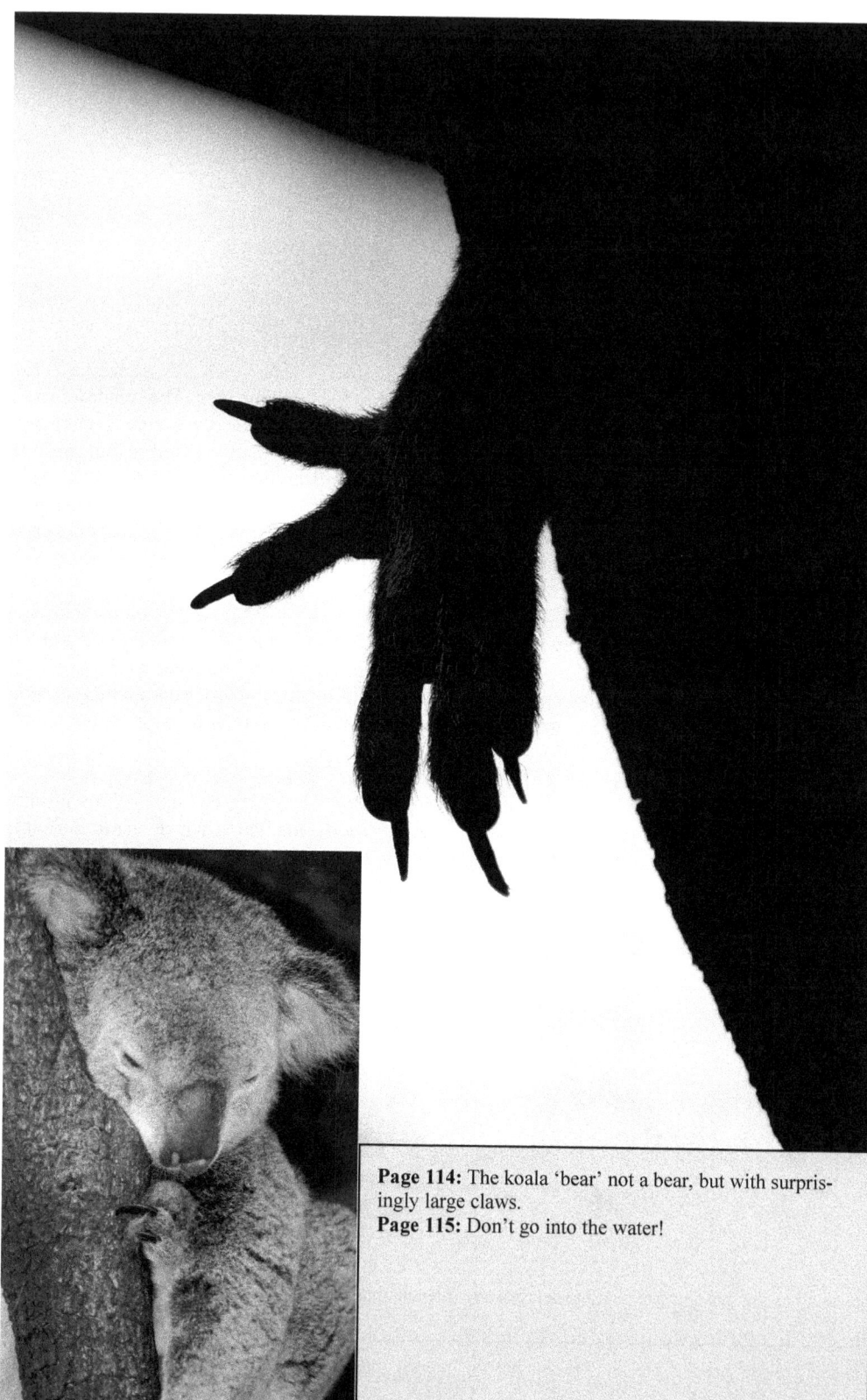

Page 114: The koala 'bear' not a bear, but with surprisingly large claws.
Page 115: Don't go into the water!

Chapter 6
The Jealous Crocodile

Thursday 13th November

After the excitement of yesterday we elected to have another day just relaxing and mooching around Port Douglas. We found an Internet café and sent a few emails around the globe whilst eating delicious ice cream. We also took some time to really appreciate the beach. The feeling of walking on the warm sand with the sun on our backs was a tonic that everyone who has missed a summer in Britain should enjoy. I really wanted to dive into the sea but there was only one small area, protected by a boom, that people could actually go into. We humans had to be protected, not from the myriad sharks out there, not from the lionfish or stonefish but the serious and silent killer that is the Northern Australian box jellyfish *(Chironex fleckeri)*. This is one of the most venomous creatures known and has caused at least 60 fatalities in Australia. It is not actually a typical jellyfish (Scyphozoa) but belongs to the Cubozoa, a group characterised by the placement of the tentacles at the four 'corners' of the body.

Known locally as 'stingers' they have transparent bodies and tentacles that can trail for many metres. They appear in shallow coastal waters (yippee!) and the mouths of coastal rivers especially after rain. The stings contain literally millions of microscopic stinging cells that discharge venom on contact with the skin. Each stinging cell will produce a large red welt and multiple stings become manifest in long lines on the body. Permanent scarring can occur.

The nets are not foolproof, however, and a large notice declaring no liability from the local council was prominent. You took your own chances here.

There were some little mangrove trees on one end of the beach and it was here that something darted quickly over my feet, literally skimming across the water at great speed. It rested on a large rock about three metres in front of me. Was it a fish? Was it a lizard? It had huge bulbous eyes set on top of its head and what looked like fins underneath and down its back. Its tail tapered to a sharp end and it was very well camouflaged. It was *Periophthalmus argentilineatus* - better known as the common mudskipper.

I was delighted to see this handsome little fellow as I had only ever seen them on TV before. I took some photographs from a distance and slowly moved in to get close-ups. I seemed to gain its confidence as it let me get close so I took the photos I wanted and thought I would get a real close-up to finish off. The little chap had other ideas and hopped across the water at lightning speed and I laughed aloud to see the comical way in which it fled.

An English couple who were photographing the sunset came across to ask what I was snapping away at so I took them to where the mudskipper was now resting and they were shocked to see it as they never knew such things existed. Nature is wonderful, especially when shared.

I went back to the apartments and we ate a hearty meal, and afterwards I asked Sue if she would come out and join me in the brush around the hotel grounds to see what we could find. She readily agreed and we set off. No sooner had we gone into the trees than we stopped dead in our tracks on hearing a loud hissing sound.

"Don't move," I whispered as I thought that this was the moment that we had met a death adder and I wanted to get my bearings.

Another hiss and I flicked the torchlight to where the noise was coming from. There, as still as the night air around us, stood a most beautiful, beguiling and unexpected animal. It was a bush stone-curlew *(Burhinus grallarius)* and we slowly backed away to save stressing it. I managed

Page 117 top: A spectacular view of four mile beach.
Page 117 bottom: Relaxing on a 'day off'.

A Daintree Diary

to get an acceptable photograph before we slipped away. It was a beautiful find and we were very lucky to have seen the bird. To be honest I was a tad disappointed that it wasn't a death adder but Sue would have another opinion. Later when I loaded the picture onto my GETAC B300 laptop I noticed a little white-streaked shape a foot or two behind the bird. It was a chick! The 'it' became a 'she' and the little one was being fiercely protected by the mother, hissing at us to stay away. It certainly worked. Nice one stone-curlew – way to go!

Friday 14th November

The name 'wolf spider' often sends a shiver down the spine of a good many people. It conjures up visions of large, hairy, fanged monsters ready to pounce and consume prey at any given moment.

Wolf spiders are in fact wonderful creatures. They are Lycosids and generally roam on the lookout for their food and in Australia they prefer hot leafy terrain such as woodlands and gardens. There are many species. You will have seen these spiders; they carry their egg sac and resulting young on their abdomens (the back end). They have large front eyes that readily reflect torchlight at night, which certainly helped me to find them. As I shone my torchlight across the leafy floor I would pick up a little glint of gold shining back of me. The floor was lit up with these jewels of the forest. Apparently their bite is harmful to dogs and cats, and at least two species prey on young cane toads. In northern Queensland a *Lycosa lapidosa* was noted biting a large toad on the head and the toad died one hour later.

Page 118: The stone curlew and chick (to the right).
Page 119 top: Sue and Angie
Page 119 bottom: The nutcracker (sweet)

Page 120 top: A wolf spider blends in with the sand **Page 120 bottom:** This crab was smaller than my fingernail **Page 121 top:** Camouflage saves lives **Page 121 bottom:** Are those *people* on 'my' beach? **Page 122:** Life springs eternal, whatever the habitat.

A Daintree Diary

I thought wolf spiders lived exclusively in leafy habitats but today I made some exciting observations on the beach. Whilst Sue was relaxing I thought I would have a look amongst the fallen palm leaves behind her and to my astonishment I saw a large wolf spider about an inch and a half long. It was silver and fawn, the same colour as the sand, and had little black dots on the abdomen. I then found another, and another. It seems that just a metre from the heads of beach-lovers were large spiders so maybe it is true that wherever you are you are rarely more than a metre away from a spider – except Antarctica when none live to our knowledge.

What a surprise…I had my camera with me (of course) and managed to get some illustrative shots of how animals camouflage themselves to blend in with their environment. Just a few metres behind the head of the beach, in the forested area, the wolf spiders were brown and darker, just like the leaf matter. How remarkable evolution is.

Just then there was a quick movement by my feet; it was a tiny crab only millimetres wide. It was gorgeous and I put my finger close to it. Each time I came close it sped sideways and I let it go with a chuckle. Why do crabs walk sideways, anyway? It seems absurd to me. They don't take their food that way so I wonder what the benefits are. I recalled that funny joke about the man who walked into a fish 'n' chip shop and asked if they served crabs. "We serve anybody, sir," was the reply. We enjoyed a long walk on the beach, finding an array of fallen coconuts and other fruits and seeds. We were so happy to have some respite from the maddening rat race and these were moments to be savoured. In addition, we were excited at the prospect of what was to come that evening. We had booked an evening trip to sail in a small boat down the Daintree River and it promised to be spectacular.

Spectacular it was: our guide for the evening was an upstanding local chap called Clay and he led us to the boat where his mate, Mick, would skipper us out on the river. The sky was turning from inky blue to black and we could make out the shapes of giant bats above our heads flying to destinations unknown. We could hear noises: a splash here; crickets there and a bird screeching somewhere in the distance.

There were only a handful of us as we navigated to the middle of the river. Mick was a mine of information, but then he had been doing the job for many years and had an encyclopaedic knowledge of the area. He knew the plants and animals not only on the Daintree but from a much wider natural history perspective. Specifically, however, we were searching for crocodiles. Mick knew where to find them and he told us about Albert and Fang. They both wanted to court Elizabeth and Fang used to be in sole charge until challenged by Albert who was desperate for the territory. Albert was the young croc and the fight between young and old was imminent. Albert won and now rules a territory of some ten kilometres, and pops in to check on his ladies fairly regularly; a sort of reptilian pimp, if you will.

He has swum right up to the boat before to warn Mick to 'bugger off' and leave his girls alone. This is particularly common during mating season. Albert tries to force him upstream and there is often a psychological battle of wills. Albert would like nothing more than to catch Mick off guard one night – it would only take one second and he would drag him under to a watery and certain death. Having said that, Mick seemed a wily old fox too and I wouldn't bet against him seeing the croc off. I recall again the different perspective I had of the sky, the moon and the stars from the middle of that quiet river. There had been a tide change and animals were stirring all around us. Small micro-bats joined the flying foxes and shot past our ears at high velocity. I have stated before that the bats spread seeds for many kilometres and are an essential part of the reproduction cycle for plants but I asked Mick how they knew which trees to fly to as some were many miles away. The knowledge of these trees is passed on by generations of bats. Mother has her young clinging to her for a few months and they fly together. The bats know which trees have fruited and even which ones are about to fruit. It's miraculous. Mick covered a variety of subjects from beautiful trees that had the most wonderful fragrance more profound than anything found in a bottle in Paris, to the root systems of mangroves and other trees. He described how these plants breathe and showed us the elaborate bird's nests hanging in them, disguised from the predatory butcherbird.

Clay told him I was a spider enthusiast and Mick told me I was in for a treat. He knew the exact spot on the side of the river downstream where a colossal web was home to large and numerous tent spiders. He kindly took a detour and before long I was at the end of the boat leaning into the edge of the complex web with my lens. I assumed that Mick was keeping a lookout for the Albert as I leaned well out of the boat and focused on the web's occupants. There were some big spiders in there but smack in the middle of the web was the matriarch, the mother of all tent spiders. She had a body length the size of a fifty pence piece and thick stout legs. Her abdomen was white with a broad brown stripe down the middle and the legs were brown too. Checking the reference books I am fairly sure that what stared back into my lens was a *Cyrtophora* sp. of some sort. She was stunning in form and stature. A formidable spider that was the paragon of arachnids and stood for everything I love about her and her kind. Nothing lasts forever, however, and I knew that it might take a second for this majestic specimen to be taken by a bat or a bird. I bade her farewell, and was grateful for the simple moment we shared on that most memorable of evenings.

Australian law dictates that you are not allowed to get within ten metres of a crocodile and Mick, ever the consummate professional, ensured that this was so. I recall thinking two things here. Firstly you would have to be barking mad to want to get that close and secondly, I assumed there was no law prohibiting a crocodile from getting within ten metres of you. We did get pretty close to one and she was incredibly difficult to spot. Larger crocs could be seen in the torchlight in the distance but when we got close they submerged and disappeared. It was quite easy to spot those red eyes in the water but I made sure I never leant too far out all the same. I wanted some stories to tell when I got back home, but not the one about the croc that swam off with my arm. Those were not the 'snaps' I yearned for. In a heartbeat the boat trip was over and we moored ready to go off for a rainforest lodge meal before a brief night walk at the Silky Oak Lodge at the Mossman Gorge. On the menu this evening was barramundi and it tasted delicious. I shall give more detail about the barramundi later as I had a particular reason not to forget this legendary fish. We shared our meal with a curious couple from Melbourne who had come to the Daintree area for a sunshine holiday. I learned that there were constant water restrictions where they lived and life there was particularly hard.

Presently, we met a few more people and all followed Clay into the forest for the tour. Torches

Page 125: This was about as near as I wanted to get…
Page 126 top: Down came a (large) spider
Page 126 bottom: It's easy to spot crocodiles at night.
Page 127: Clay Mitchell – in search of the whistling spider
Page 129: Mygalomorph spider found in vertical clay bank

A Daintree Diary

in hand, we were able to spot nightlife and we quickly learned not to shine a torch directly into the eyes of an animal, but near to it so as to minimise any stress. Clay's torch was huge (well, he was the leader) and he could pick out foreboding fruit bats with a one-metre wingspan, circumnavigating tree tops from many metres away.

As we walked along I was at the rear of the group but Clay suddenly exclaimed "Carl, have a look at this" and I sped to the front in time to see that a large huntsman spider had lowered itself from a tree right in front of his face. Some of the more nervy tourists reeled back in expectation of some kind of disaster but Clay calmly let the spider land on his upturned wrist and walk up his arm. Most people relaxed as we both explained that the spider meant no harm – we were in its territory after all. The value of education in the field cannot be underestimated.

We continued on to see small marsupials, a white tailed rat, and many insects including katydids and dragonflies at perfect rest on the end of leaves. A bandicoot was on the move in front of us and he disappeared in a flash. I felt a little guilty, actually, because although Clay was the undoubted leader and expert, several of the group seemed to be following me. I kept deliberately stepping of the trail to find my own insects etc. but people kept coming with me and photographing the same things that I did. I am sure that Clay knew this but was very patient and very professional about it. One chap stayed right with me and every time I found something he would say "Awesome…what is it?" I had to laugh.

Just before the end of this walk, where some refreshing rain was falling I happened to shine my torchlight onto a tree to my left. On its bark rested a huge huntsman – the largest I had seen so far. This was the size of my hand span and it was a mature male. I was unable to tell

from this specimen if it was *Typostola barbata* (the giant green huntsman) or *Halconia immanis* (the grey huntsman spider) as it was reddish brown in appearance and one cannot usually tell the species of a spider by colour alone. These spiders get mistaken for tarantulas but they are certainly not. They are large, yes, but also timid and do not bite readily unless provoked. Huntsman spiders get confused with wandering spiders. There are wandering spiders in Brazil that are dangerous to man. Indeed, *Phoneutria nigriventor* is notorious not only for the toxicity of its venom but that it is not at all reluctant to attack humans. I have seen a stunning photograph of a Brazilian barber literally trying to fend one off at his shop doorway with a stout broom. Aside from the intense pain it is said that the venom can cause priapism – uncomfortable erections that can last for many hours. I am unable to ascertain what happens when a woman is bitten! (EDITOR'S NOTE: It causes a long, uncomfortable erection of the clitoris, also known as clitorism).

Many people are curious as to how I am able to tell if a spider is male or female. The stock answer to this is that it is best done scientifically. Usually you examine the shed skin of a spider, but there are some other characteristics that are dead giveaways. For example, the spider I was looking at had long, thin legs and a thin abdomen, but most of all it had developed sperm sacs on its pedipalps. These are the two little 'extra legs' on the front of the spider that look like the creature has little boxing gloves on. I could go into the wonders of spider taxonomy and discuss the many attributes of the slit sensillum, the uterus externus, and the gonopore slit but I shall refrain on this occasion. Now malphygian tubes really do get me excited and I often used to debate the pros and cons of such a useful part of the anatomy with my very good friend and wildlife expert, Mark Titterton. I am way ahead of the anoraks you know – I even have a diploma in sexing juvenile Theraphosids (tarantulas), which I am fiercely proud of.

On the drive home we came across a dead wombat on the side of the road, which saddened us all. This weird-looking animal is neither a 'wom' nor a bat. It is in fact a marsupial. They look like little brown bears with stout legs and have very sharp rat-like teeth and claws to dig out burrows. We were witnessing another senseless death on the roads. You wonder with all that space out there why car meets wombat with sickening regularity. When we returned home I became unable to resist another short walk outside and found more wolf spiders and a large nursery web of juvenile spiders awaiting their first excursion into destiny.

Saturday 15th November

Susan never slept well due to the searing pain of the sunburn. It looked raw and angry and I was very concerned for her health. Therefore it was sensible to opt for another quiet day by the pool, reading and revitalising spirit and body. I wrote up my notes from the previous days and Angie decided that the pool was a good idea too. She did win the award for the quote of the day. When the heel started to come off the bottom of one of her favourite pairs of shoes she sang aloud "You picked a fine time to leave me loose heel." I told you that Liverpudlians and comedy were bed-fellows. Nicole from reception gave us some of her lunch in the form of lychees, which had great restorative properties. She was always willing to help and indeed there was not a member of staff or management who didn't. I could ask for a new set of pillows or a large cane toad and they would deliver the goods – now that's service.

In the afternoon we all took a stroll into town to buy a few gifts. I also booked a car for tomorrow as I would be travelling north to Cooper Creek for another day of magic and mystery. Word seemed to have spread around the little town that there was a mad Englishman around with an unhealthy fascination for spiders but I was always made welcome by those who spoke to me or helped me organise private trips. Sue and I dined alone in town in the evening. There is a relaxing place called the Iron Bar and it's here that I couldn't help but overhear the man on the table next to us. Well, it wasn't so much overhearing as him talking excessively loudly, but he was an altogether fascinating character. He was a wiry chap with khaki shirt and pants, a long 'Catweazle' beard and a stupendous hat, which suited him. When he took it off it revealed a bald Kopf, which completely changed the mystique of the man. He was a true character, though, and obviously a regular at the bar. I was to see him around town a few times. The young waiter came to take his order, which he gave and finished off (in best Aussie accent) with the comment "and throw a few shrimps in for free mate." He was talking about his student days in Brisbane, including the fact that he was always completely hammered, and something about "I had to shine a torch under the blanket just so I could see her fucking arse", which creased me up. The guy was a social hand-grenade but I really liked him and wished I could have spent an evening listening to his stories. He was loud but good-natured unlike some of the tools who want a fight after a couple of schooners.

On our way back from the restaurant we decided to go up a steep hill at the back of the main drag and walk along that road. I wanted to look into the banks to see if there were any spider holes. I had done this several days before with no luck but gave out a whoop as I found a burrow about the size of a ten pence piece in the side of the bank. What's more there were a couple of thick, hairless, black legs protruding from it. In a flash I got a piece of twig and 'tickled' the entrance, and the spider shot out and bit it straight away, then darted back in.

"We've got a feisty one here," I said as I inserted the twig a little into the hole. Suddenly a third of it disappeared down the chamber as the spider put up an almighty fight. I was impressed – very impressed and gently pulled the twig out with the black beauty on the end of it. I am sure it was female by the size and shape and she was jet-black all over, and much bigger than a house spider.

It was a Mygalomorph, which means it had forward-striking fangs that operate in a downward movement, whereas other spiders, called Araneomorphs (or true spiders), have fangs that move from side to side like pincers. I photographed the glorious creature at the burrow entrance as the rain began to come down so we called it a night and went home. I used to keep some 1500 tarantulas at home at one time but I now have such fun finding them and photographing them in their own habitat. There is a sign in a nature reserve near my house that says

TAKE only photographs, LEAVE only footprints, KILL only time.

I tend to agree.

The tropical rainforest at Cooper Creek

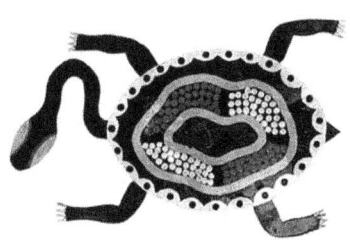

Chapter 7
The Ghost in the Forest

Sunday 16th November

And so it was that I found myself driving north to the Cooper Creek wilderness, only a few kilometres south of Cape Tribulation. The topography was different from the moment I crossed the Daintree River on the ferry. The plants seemed more primitive and the scenery was wilder.

I was on the way to meet Neil and Prue Hewett who owned some seriously unique land and the word was they were rightly fiercely protective of it. I also heard that they were the people to introduce you to an unforgettable world and share their knowledge of the area so it was with this in mind that I pressed the accelerator and twisted this way and that around the chicanes to my destination.

Several days earlier Neil and I had a telephone conversation where I explained that I was looking for the whistling spider and really needed to find one before I went home. He informed me that he knew exactly where there was one on the grounds and that I was welcome to come and see it. I will explain more later on about the nature of the whistling spider and why I so wanted to find it.

I drew up on the driveway where Prue was already standing. We exchanged pleasantries and Neil arrived to give me a tour of the area. He was due to lead a group of tourists (I guess I was one too) that evening but until then it was just he and I.

Neil had lived for many years with Aborigines learning his trade and always walked around barefoot. He was a fount of knowledge and I immediately warmed to him for the way in which he spoke about wildlife and the environment. This is how he describes his tours on the official leaflet. 'The secret places of the world are difficult to find. Sought by many, found by few, these amazing treasures are concealed by a timeless barrier of protection. Cooper Creek wilderness contains an awesome expression of diversity and perseverance.

Access cannot be achieved from the convenience of public boardwalks; it requires far greater determination. Treks are on the forest floor, unspoilt by artificial structures.'

Every word is true. This is a place where I might place a foot where no white man has trodden before me. It has huge fan palms and thick vines and there were more animals here than I had previously seen.

After only about ten paces I managed to take the most exciting photograph of the whole trip, thanks to Neil's knowledge of every tree, every nook and cranny.

He pointed to a tree that had green and silver lichen on it. "What do you see?" he asked and I looked very closely.

I couldn't see anything.

"It's right in front of your eyes," he said with a knowing chuckle.

Suddenly, as if by magic a shape formed in front of me and there was the most sublime lichen spider I had ever seen; maybe even the first live specimen I had ever set eyes on. After checking the literature in my own library I believe it to have been *Pandercetes gracilis*. I carefully photographed it whilst not exhaling for fear that the beastie might disappear in a puff of smoke and then I exclaimed "Yep, got it: the spider."

Neil replied "Is that all you see Carl?" to which I replied yes.

He pointed an inch above the spider and there clustered all around the mother were the ghostly translucent figures of hundreds of her young. They were almost invisible against the bark, only identifiable through thin traces of body shape and black, haunting eyes.

My jaw literally dropped. I would truly have been happy to go back to England with this one image in my mind – nothing else could better it. I was privileged in the extreme to have seen this and I thought of absent friends who would love to have stood at that precise spot.

Neil would find more of these ghosts in the forest as we strolled around and I was only able to find the occasional one or two, so good was the camouflage.

As we approached a large corrugated shed he said that the whistling spider was housed within. Excitement mounted. He lifted up a blue plastic box and underneath it was webbing but the occupant could not be found. I looked on the bottom of the box and there in front of my eyes was a huge female huntsman spider guarding an egg sac.

Page 133: A lichen spider. Can you see it?
Page 134: If you didn't see it, would a predator?
Page 135 top: A female huntsman spider guards her eggs
Page 135 bottom: The legendary whistling spider – this one at Cooper Creek.

Now, that was a precious find in itself and a pleasure to see.

"Over here," Neil said as he lifted up some corrugated tin to reveal a small brown spider about three inches long. "That's it; that's your whistling spider," he said and I admit to being elated and disappointed.

I had hoped for a larger specimen but considered myself lucky to have found one. It was to all intents and purposes 'just' a brown spider but it was fat and well fed, which was unusual for arachnids found in the wild. Suddenly a large centipede that had been coiled by the spider shot off to hide elsewhere. Both Neil and I commented that it is a strange phenomenon that often when you find spiders, centipedes are never too far way. Why is that?

We then walked around the orchard and this is where I whiled away a couple of glorious hours finding specimens that I had never seen alive before.

I saw for the first time a species of bird dropping spider *(Celaenia excavata)*. This beauty mimics a bird dropping for camouflage and what's even more interesting is that it serves as a

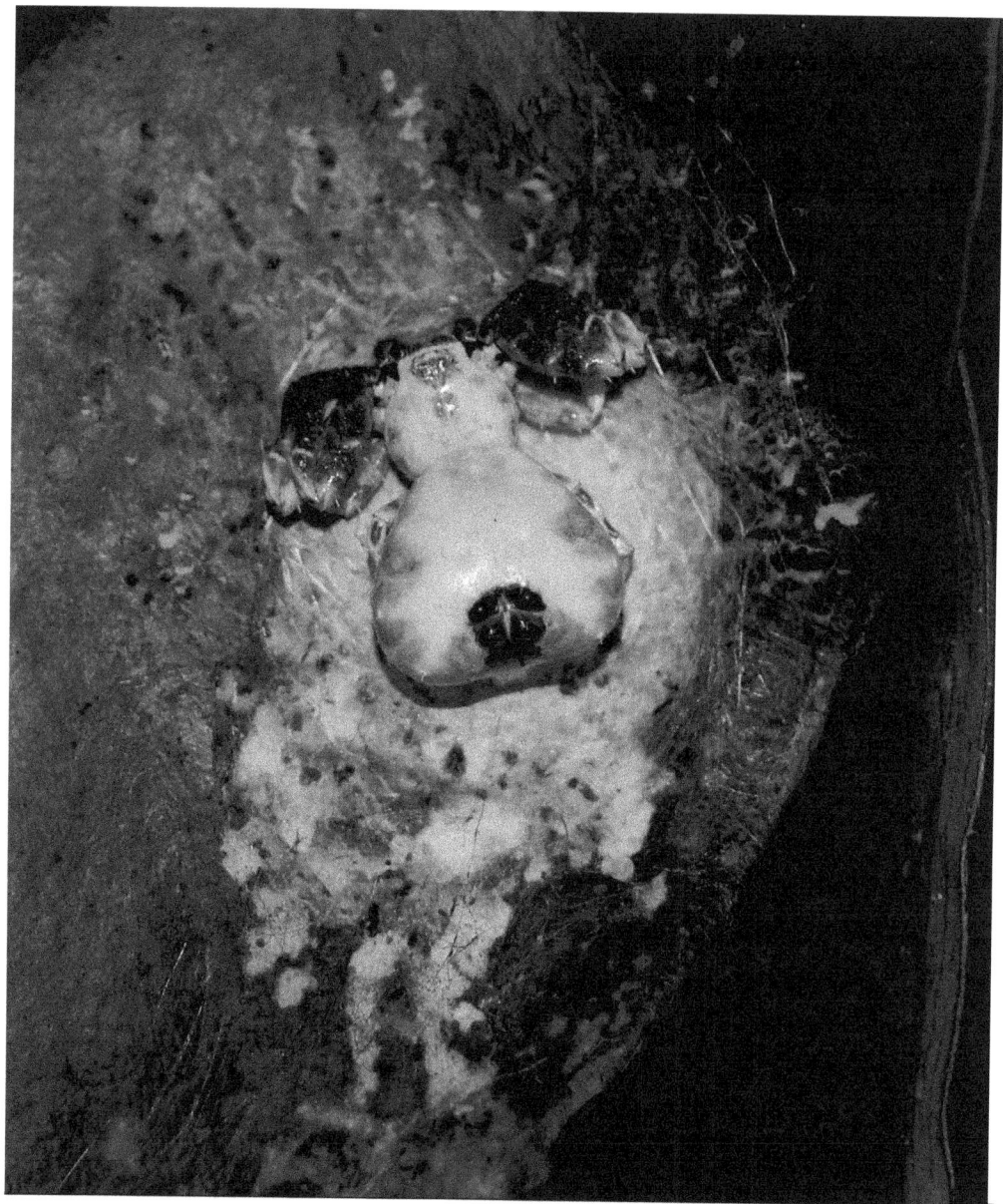

Page 136: A bird dropping spider – yet another of nature ruses
Page 137: Boyd's forest dragon *Hypsilurus boydii*
Page 138 top: A big black spider attacks me and defends her young
Page 138 bottom: A simple orchard – but so much life on those trees.
Page 139: A St. Andrew's cross spider.

defence (by ensuring other animals believe it to be only a bird dropping and not of interest to eat) as well as an attack as prey walks close by believing that it is just a safe bird dropping. There were St. Andrew's cross spiders *(Argiope* sp.*)*, jet-black mygalomorphs with young raft spiders, *Nephina*, jumping spiders, opilones and massive huntsmen, not to mention all the large stick insects and praying mantids. It was a veritable shop window chock-full of arachnids and I for one had my nose pressed hard against the glass.

Neil told me that one day he found a large spider that he never recognised in one of the trees. He rang an eminent arachnologist at the Queensland museum only to be rudely told not to bother him and that "If he (the arachnologist) saw an elephant in a tree he wouldn't ring him would he?"

This didn't surprise me. There are certain 'intelligent' arachnologists along with other experts that publicly declare that everyone – professional and amateur – should share knowledge and work together for the good of the spider when in reality they are nothing more than neo-Stalinist control freaks who believe that their view is the only view and that everyone else is a fool. I like to think of them as clever idiots. Don't get me wrong; most are gregarious, genuine people who cherish a deep love of the subject and willingly share information with us mortals. However, so-called experts can describe to science as many spiders as they like; they can pull as many apart and count the hairs on the spider's leg or how many cheliceral teeth it has but when the day ends, what have they done to protect the species? It's the kind of elitist attitude that this 'expert' displayed that puts people off wanting to help.

But don't let it get you down folks. Go your own way, believe in yourself and keep learning from your own experiences. It's your world too.

Amateur naturalists can learn many things that so-called experts miss. Neil called me from a few metres away and showed me the most remarkable insect I think I have ever seen. It was a katydid about the size of my hand and nestled tightly against a tree. It was like something out of a science fiction film, with bulbous eyes and jagged green and white spines on its legs. I would never have found it without Neil and I would challenge others to have done so. The best I can find is that it is called the spiny-legged rainforest katydid, which is not very creative but very apt, indeed.

Neil's children were very efficient at finding spiders for me. They knew the territory of course and they had many questions for me about England. They were wonderful, cheerful children of which Neil and Prue will be rightly proud.

As we passed a tree Prue pointed to a little brown and green plant growing out of the bark. It just looked like shoots of some kind but this was no ordinary plant.

Oberonia titania, also known as the royal fairy orchid, was discovered by Prue's granddaughter Tojah in 2000 when she was 11 and liked to climb the fruit trees in the orchard. There are several specimens on rambutan trees.

They sent an image and the name to their friend, environmental scientist Andrew Small, who added it to his plant list on the web. He was contacted by an orchid specialist who queried the listing because the plant was believed to have been extinct from 1960. So, basically word of mouth makes it rare and interesting, but Prue is yet to find any written declaration.

In 2007 Prue found a relative, *Oberonia meulleriana*, in the orchard. This indicates the possibility of many rare species existing in the forest where we can't see them; maybe up in the canopy. The orchard has become the recipient of many orchids and ferns as the rainforest takes back its land from the top, down! We had a brief dip into prime rainforest where I saw the biggest *Nephila* spider of my life about fifteen feet in the air suspended on a web that was built between two trees at least three metres apart. No wonder they have a reputation for catching birds and bats. Those webs are strong and the spider will take anything it can overpower.

It was time to get to the start point for the real rainforest trek. On our way back we went into a small outbuilding and Neil showed me a bright green white-lipped frog on one of the windows,

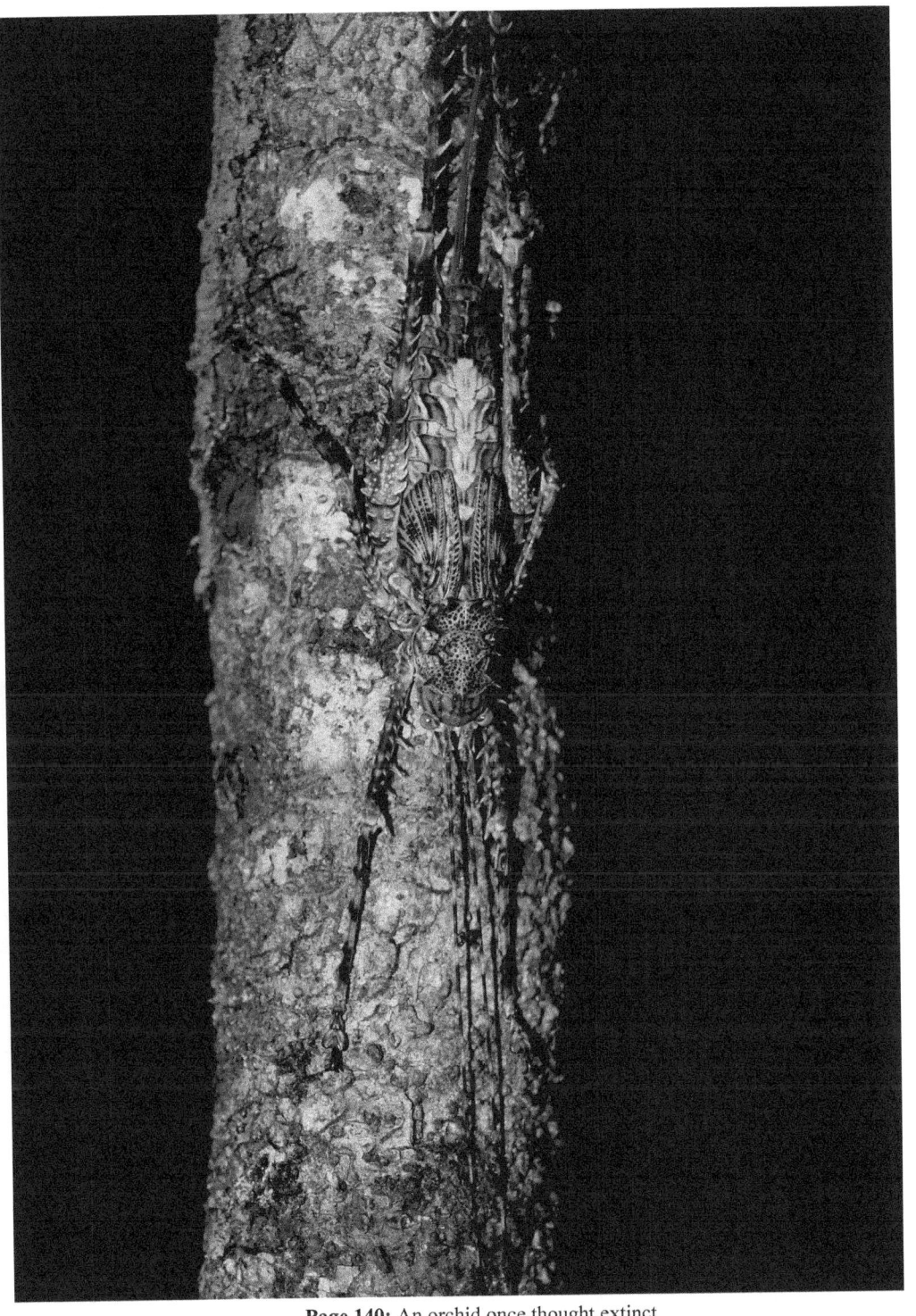

Page 140: An orchid once thought extinct
Page 141: A huge katydid almost invisible on tree bark
Page 142: A charming stick insect at Cooper Creek
Page 143: A *nephila* spider the size of an adult hand.

and two giant huntsmen just above our heads. We were both having enormous fun.

We gathered at the meeting point and a group of Germans, Americans, French, English (me) and Australians headed off. Neil was very careful to explain that smoking was prohibited (the message eventually got through to the Frenchman) and that photography was generally not allowed unless Neil agreed. Torchlight must not under any circumstances be shone directly into the eyes of any animal and people must spread out in single file. Each person would have ten minutes at the front with Neil, then go to the back of the line so the next person could benefit from being close to Neil as he talked and located animals and plants. In my opinion he was quite right to be so protective. This is one of the last areas of its type on earth and must not be damaged or abused in any way.

I began the walk at the back of the group to give everyone else a chance but within a couple of minutes Neil had shouted me to come up and look at something with him. I walked past the people staring at a fixed point and arrived at a large orb web of several feet, with a stupendous *Nephila* spider in the middle. It had red legs with black-banded segments at all the joints, a black carapace and a creamy abdomen. It was the size of my hand and absolutely gorgeous! Imagine one of those orb web garden spiders in your back garden, but the size of your hand. Excited? Someone asked me if it would bite and I said I wouldn't fancy putting my hand too near it. It's strange that that is one of the first things people ask about spiders. They never ask 'what would happen if there were no more spiders left on earth?'

As the trek went on we saw a white-tailed rat and a fawn-footed Melomys, which is a rodent. There were many more interesting spiders and frogs, and a charming Boyd's forest dragon. This forest was so untouched by man that the animals had not yet developed a fear of us. This was demonstrated as we walked past a large leaf extending from a tree where a little bird was blissfully sleeping. The same was observed with dragonflies.

As we walked on it became clear that the three young German lads were having a bit of a lark and not really interested in the rainforest. They were speaking their native tongue and laughing and hanging on vines when at the back of the group. Little did they know that I speak reasonable German and suddenly exclaimed "Leiser Mensch!" They looked at me, astonished. Then we got talking in German and they explained that they were on a round-the-world trip. They were good lads and I wished them well but I told them jokingly that they wouldn't get out of this forest alive if they didn't hush and stop scaring any animals that we might come across. We had paid for this, after all. As we walked on, one of them did have very good cause to shout.

"Schlange! Schlange!" he said as a black snake approached. Neil appeared almost at once and confirmed it was a small-eyed snake *(Cryptophis nigrescens)* and that it was potentially dangerous. He asserted that we should all remain calm and let it continue on its way, which we did. I later looked the snake up and it is indeed dangerous, possessing a myotoxic venom. It preys on skinks, lizards and geckos, and only grows to about one metre.

I had been looking for scorpions the whole trip and found nothing. Neil assured me that he knew a tree where they resided and sure enough, we arrived at said tree. Shining a torch would just have scared the little scorpions into disappearing again behind cracks and crevices in the bark so Neil turned to me and said "It's time - have you got it?"

I had something new to offer Neil's tour. When I have searched for scorpions in the past I have used an ultra violet light, or blacklight. Scorpions actually glow in the dark when exposed to UV and are very easy to spot. Neil had been aware that there were some scorpions in this tree but he did not have any idea how many. I flicked the blacklight on and the scorpions lit up like lights on a Christmas tree. They were everywhere: on the bark, on the roots, on the floor in front of us and as I swung around, they were on other trees too. The gathered throng gasped as one and I know that Neil was impressed; so much so that he will now be incorporating this neat trick as part of his tour.

As the tour ended Neil reminded us just how lucky we all were. We had just walked in rich and diverse prime tropical rainforest. It is some of the last of its type on earth and we were amongst a significant minority of humans that had done so as not even many Australians did this. It is an ancient and spiritual forest and I find it difficult to describe how I felt as I stood looking up at trees that were 1500 years old. The words 'humbled' and 'insignificant' come to mind.

The party left and Neil fetched me some water before I drove back. I needed to micturate and Neil pointed me down a short path into the forest where a corrugated 'dunny' was situated. He said "If you are lucky you will find a spider in there," and as I opened the little door, sure enough there was a decent-sized huntsman at eye level. I smiled - it had been great fun. Even when I rejoined him (after tripping over a bulky cane toad) he pointed to a large longhorn beetle: jet-black with ten red spots on it; sitting on the wall of his house.

We said our goodbyes and as I drove back I realised that so much of the wildlife I had seen had camouflage as the prime defence. Don't get seen. If you don't get seen you won't get eaten. No sudden movements because if you move, you're more likely to be dinner unless you are at the top of the food chain.

Even though I am back in middle England I can live my life knowing that such an important place exists on Planet Earth. I can visualise the scenes over and over and hear the sounds in my head. I am indebted to Neil and Prue, and as custodians of Cooper Creek they are a credit to the human race.

Monday 17th November

Martin Luther King may have had a dream but so did I. I have never fished, and I always wanted to have a go and catch a huge blue marlin. That wasn't going to happen in this area but Sue and I did book an afternoon trip so I had the opportunity at last to catch something and we were really looking forward to it.

This wasn't any old fishing trip, mind. This was a trip with the one and only Bill Clarke. This was a man who, once met, would leave an indelible impression on your soul. Some might say he was a 'typical' Aussie full of bravado and stories of the outback. Others would say he was egotistical and self-centred. Others would say that he is a fascinating character who makes you think, no matter what the opinion or subject. I am in the latter camp. It would be easy to fall out with Bill at the drop of a hat but he was (and is) one of the most genuine men I have ever met.

I guess he was over six feet tall, and had the trademark cigarette in the side of his mouth, a shock of wavy hair and the physique of a bodybuilder – he was the man to be out on the water with. He has been doing this for over fifteen years and seen many competitors and dreamers come and go. He knew this river, all right, but most of all he knew the fish. There were only four of us, and Bill. As we set off we made our introductions with the other couple who were retirees from England.

Bill began by saying that there was no bait on board and if we were to have fun we didn't want piddly little fish anyway. Our bait was going to be the size of the fish you might catch back in the UK so imagine, he said, the size of the buggers coming to get it.

There had been much rain the day before and a high tide, which affected the behaviour of the bait (I think it was mullet) that Bill required. After an hour casting his net over the side we had nothing so he said "Right – we are going to my special place where I guarantee we will get some bait," and we sped off with anticipation. We arrived in a little gully and Bill turned the motor off and asked me to jump out and tie it to the tree on the bank. He asked the group to stay and said that he and I would not be long. He warned that there were crocodiles about and that the group should stay put. We left the puzzled threesome in the boat and I followed Bill over some very slippery ground; an open clearing leading to another little river edge.

"Listen mate, I am not kidding; there is a huge croc around here and I don't want you to wander. Stay with me and don't ever, ever turn your back to the river." He wasn't joking, and this was no game. "He was here this morning."

I asked him how he knew.

"Look there, mate. Can you see where he has dragged his body from this side of the track to that side? Look how wide that is – he is big and he is here somewhere."

I looked around nervously, scanning every bit of bank for several metres but these reptiles are brilliant at hiding and it would be hard to spot him at the best of times – even one so large. I thought I would take a photo of the mark the croc had left in the track and I took a few paces towards it.

Bill went mad: "Fucking hell mate! I am serious. Keep away from that track; he is here somewhere and he will have you. If he comes for you, you won't stand a chance on this slippery mud, which is just what he wants so back away right now mate."

He was not happy and that would not be the first time today, either. He was, however, very justified and had only my safety in mind. Over the years he had learned of people being taken by crocs and it was all too quick. He didn't want me, a curious Pommie, being the latest casualty.

He knew from a previous comment that I was a spider enthusiast and he had stated that I had a treat in store. On the way to get the fish he took me aside to a huge web in the bushes where several large arachnids were gathered together. They were impressive and quite hard to photograph as they were located several feet up. I did lean in and manage to get a face full of web, which triggered immediate interest from the spiders. They turned and began to run down towards my face to investigate so I pulled away to let them get on with repairing the unfortunate damage that I had done to their accommodation. I would have liked to have spent some time here but there were other pressing matters.

We walked to a little gap at the water's edge and he cast his small fishing net, asking me to stand on the bank but be wary at all times. I did as he said and he started throwing fish at me to store in the bucket I was carrying. The fish were mainly mullet of varying sizes but we had the odd prawn too. When we had thirty or so we set off back to the boat and the trio was still there, roasting under the hot sun. Sue and I wondered upon our return back home, what would have happened if the croc had gone for them.

Bill took us to a location where he thought we would catch fish and we all had a rod each, set

Page 147 top: The one and only Bill Clarke
Page 147 bottom: This spider shone like a jewel against the river bank
Page 148: The large catfish (Arius sp.) that Susan caught.

the bait on and plonked it in the water; then we sat back and did what fishermen do – waited. "Welcome to my office," said Bill as we looked around at the river, the mountains, the forests and the eagle flying high in the distance in the clear blue sky. "You can keep your offices wherever it is you work; look at my office," he repeated. I felt compelled to respond since he was obviously keen to engage in more sport than fishing. It was Bill's way of testing the metal of the crew!

"This might be your office Bill, but you have to come into work today – whilst I am on holiday."

He agreed it was a fair point and went on to tell us about his life and his ex-wife in particular. She was French and gorgeous by all accounts. In the end they split up because he tried living in Paris and couldn't and she tried living in Australia and couldn't. He still loved her deeply, that much was patently obvious. "She was a ripper," he would say.

"Check that line," he volleyed as Susan's line bent and she picked it up in a flash. Bill advised keeping the line on the water and letting the fish run a bit then bringing up the rod sharply and reeling it in. Sue did this to perfection and out of the water came a very large catfish. Sue was elated and I was genuinely pleased for her and photographically recorded the occasion. Bill took the hook out and threw the fish back. He never takes any fish and he always uses special hooks that dissolve over time.

I sat back believing that my turn would come. After a while we had not had another bite and Bill reckoned it was unlikely in that spot so we went to another area and sat back again. A pretty little swallow landed on the front of the boat and twittered merrily. This was being at one with nature: no lorries, no children chattering, no shouting, no aircraft, no loud music, no one selling *The Big Issue* and no TV adverts. All I could hear were the thoughts inside my head, and it was bliss.

Bill continued with his story. He hated Europe and the weather and the crowds. He said he was not going to send his kids to Paris this Christmas time to be wearing "thick coats and duct tape." He had an opinion on everything. He deeply disliked art and the pretentious hangers-on pontificating over paintings with drivel such as "look at the brush strokes on that daffodil." I thought he had a good point there.

Just then, the retired lady's rod bent and she reeled in a large fish called a grunter. We were pleased for her and as she sat down and put the new bait in the water, she caught another, almost immediately. It was a superb Mangrove Jack. So the blokes had nothing so far and I was beginning to wonder if I would get anything.

Bill regaled us with the story of three youths on the Paris metro who began insulting his wife. He could not understand what they were saying and when he pressed his wife she said "something like a slut and a traitor for marrying an Englishman." Bill told her to pick the lad out and she reluctantly did so. He decked him with one punch, then when on the floor said "that is for calling me a Pommie," then he stuck the boot into his testicles and said "and that is for calling my wife a slut." He had no time for his ex-wife's family who considered him a 'mere fisherman' until he told them that he exported fish to Japan. Then he became (and he said this in a comical French accent) "an entrepreneur - what a load of bollocks" he said smiling, cigarette still dangling from the corner of his mouth, the blue smoke auditioning at the curled hair on his forehead. He gazed into the distance and his mind was haunted with images of that beautiful woman and the fabulous moments they had spent together. I was deeply saddened for him, though he wouldn't have wanted that one bit. Suddenly, my own trance was shattered when my rod bent double.

"Get that rod mate," Bill shouted as I jumped robotically to attention and fumbled for the implement. I have never even picked up rod before so this was new to me. In a matter of seconds the fish realised what was going on and powered like a rocket towards the safety of the mangroves. Meanwhile, I was trying to both let some line out and control the reel, none of which happened and Bill was going mental shouting at me "Fucking hell, he's gone. Too late, too late mate – he was really really big." I was shaking with frustration that I missed it and Bill was incandescent with rage, not at me, but for me. He so badly wanted me to catch a fish for the first time on his boat, in his company, in his country but I blew it.

There was silence for a moment and I asked him what it could have been. He explained that it could only have been a huge Mangrove Jack or a barramundi but that it really was massive. In mitigation, and Sue would agree, the strike was not the same as the other people had. Where they had a few seconds to let their fish run then pull it back, mine was bigger and more powerful, and barramundi are renowned for being able to make the mangroves before being caught.

My male pride was injured and Sue's soothing words of consolation were of no use, and she knew it. I truly was pleased that she had caught something but I was going home with a horrible empty feeling, as was the other chap who never even got a bite all afternoon. So my tale really is of the one that got away. It was huge, it was a monster, it would have taken two men to land it, it was called Moby Dick and don't tell anyone anything otherwise. The fish had won today, though and that's fair enough. May the leeches of a thousand rivers affix themselves to its scales. The fish that we used as bait would have been great as a catch as far as I am concerned. I am not saying that this experience constituted a bottle of whiskey and a revolver in the library but I have lamented the missed opportunity since that day.

Bill won't agree with me but whilst I agree that his 'office' is wonderful, my own 'office' is a fantastic place too. Walking across the snow-laden fields with my border collie on a crisp winter morning is a most pleasurable experience. I am glad that we have seasons in England, but I am admittedly not too fond of the cold, grey rainy days that seem to pervade the deepest nooks and crannies of Oxfordshire. We are both lucky to enjoy what we have and both able to experience and acknowledge each other's culture. He is quite right to be fiercely proud of his country and I am a proud Englishman with a love of my birthplace too. As far as I am concerned, to have been born English is to have won first place in the lottery of life. I would love to have a beer with Bill, and one day I am going to go back and catch my first fish on his boat with his rods and under his guidance. He's a first class skipper and I wish him a happy and prosperous life.

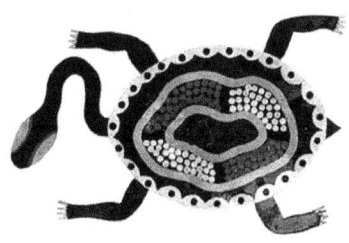

Chapter 8
Down Came a Spider

Tuesday 18th November

As the early morning sun burned its way through the blinds I awoke with the depressing knowledge that this was our penultimate day in Far North Queensland. We would all spend time together today but tonight was my very last private trip and it was a biggie.

I had contacted Clay who took us on a tour some evenings previously and he agreed to a private night tour to find the elusive whistling spider. He knew where they were and would guarantee that we could find some to photograph in their natural habitat. This was tremendous news and would cap a fine trip.

I must keep the location secret to protect the spiders, but I will say it was a very surprising one. We had to travel quite a few kilometres from Port Douglas to get there.

One of the key objectives of my trip was to find and photograph this whistling spider – the elusive tarantula that people spoke about with high regard. Having chatted to many people in shops and restaurants, and even in the street it seemed that the locals had differing ideas of what a whistling spider really was. When I showed some photos from a book people were picking out *Nephila* orb web spiders, trapdoor spiders and all sorts of different species so I had to get to a place where someone knew what they were talking about. That's where Clay came in.

I already said that I have studied and bred many species of tarantulas over the years, but not Australian ones so I was not very conversant or clear about what I was looking for. I did know that there were three types of whistling spider indigenous to Australia and finding any would suffice. These were *Selenocosmia*, *Selenotypus* and *Phlogiellus* spp. They were known simply as 'Australian tarantulas' and would most probably be found in burrows in open grassland. A bite from one of these has been known to cause severe pain, headaches, nausea and vomiting

Page 152 top: The creature I travelled 10,000 miles to find – the whistling spider
Page 152 bottom: She came to check me out.
Page 153 top: This wolf spider attacked me with no hesitation
Page 153 bottom: Trying to tickle the tarantula out of her burrow

Page 154 top: She cannot resist and exits her tunnel for a split second
Page 154 bottom: Nearby – the ubiquitous cane toad

Page 155 top: Mission accomplished – thank you my lady.
Page 155 bottom: Spider being moved back to her burrow.

for up to six hours. Such a bite is fatal to dogs and cats within one hour. When aggravated the spider produces 'whistling' or hissing sounds and they are prized pets in Europe and the USA. I should say that this hissing noise is well known amongst tarantulas across the world and it is not exclusive to Australia, which will disappoint some people. Stridulation (as it is known) is performed by spiders from Africa such as the Mombassa golden starburst baboon *(Pterinochilus murinus)*, the king baboon *(Citharischius crawshayi)* and the South American goliath bird-eater *(Theraphosa blondi)*, which is the largest species of spider on the planet. I myself have heard many of these spiders perform such a defensive sound by rubbing their chelicerae together.

Although Clay was an experienced local guide and an unquestionable authority on natural history he had never done anything quite like this. Spending an evening looking for spiders was a bit different but something that he took to with gusto. We arrived at the location: flat scrubland with plenty of trees. The floor was littered with leaves and the ground was not unlike concrete. We were on the cusp of the rainy season and all the animals seemed to be preparing for it. Some of the spiders, like the huntsman I found at Cooper Creek, had egg sacs waiting to explode with young that would feed on the abundant insect life that the rain would release.

We were looking, quite simply, for holes in the ground; probably golf-ball-sized. Clay and I split up and after twenty minutes our torchlight picked out the square sum of nothing. Would the beastie elude me? Had I come this far only to fail in finding whistling spiders in their own backyard?

Suddenly Clay called me over to a burrow. There was no sign of an occupant but young spiderlings were frantically rushing around at the entrance. I approached with my 'prodder', which in reality was an old car aerial that extended into holes and burrows to save my own fingers being bitten. It is an indispensable item of spider-hunting equipment.

We couldn't tempt any adult to come out and play so I said we should mark the burrow with a stick and return later. Clay assured me we did not require a stick as he knew where we were in relation to the tree behind us and we would find the burrow again easily. I too 'mentally' marked the tree, which had an unmistakable 'O' in the bark and we split up again. I came across some striking and unusually aggressive wolf spiders. Whilst I was flat on the ground photographing one it decided to attack me and made a dash for my face. I captured the moment perfectly on my camera. I have to give the little guy credit; he made me jump, for sure.

After some time Clay again found a hole and we approached with caution so as not to disturb the occupant. As we neared the entrance brown legs could be seen extending out of the burrow and the female (who had young) was waiting for a passing meal. I was very excited, this was something I had wanted to do for many years and here I was, at last doing some real fieldwork in Australia.

She was several inches long and not possible to identify by colour or form alone. However, it was a whistling spider and the next task was to coax her out of her home to get a good look at her before returning her unharmed.

This became a comedy. Normal practice for me is to get a piece of card and when I have coaxed the spider out with a blade of grass that feels to her like food, I block off the entrance so the arachnid cannot return. This has to be done in a split second as they rarely seem to fall for the same trick twice.

However, I had no card and neither did Clay. Undeterred, he picked up a large leaf and said he would block the spider off with that. I looked at him with incredulity and said he wouldn't be successful with such a flimsy barrier, but he was very sure.

After a few minutes the opportunity arose. The spider made a lunge for the grass and she was completely out of the safety of her burrow by about five millimetres for a split second. Clay brought the leaf down and covered the burrow entrance but the spider was having none of it. She literally punched the top of the leaf and squeezed in through the hole it had made. Clay was mightily impressed and was beginning to see that spider-hunting was not as easy as it might sound!

"Wow! That was pretty amazing," he ventured and I too admired the spider's bravery and determination in the face of two giants trying to divert it from its home.

Rain began to fall as we set off for the first burrow to see if anything had emerged. We looked, and looked…and looked. Nothing. We were definitely by the right tree but somehow the spider must have covered up the burrow with one of the thousands of leaves lying on the floor. We had to give up the search after about twenty minutes and chalk that as a 1-0 to the spider.

The last chance now was to go back to the second burrow, which we had marked with a spade and try to entice that female out. We certainly couldn't find any other burrows, which surprised Clay as he thought there would be many. I reminded him again that finding tarantulas is seldom an easy task.

We returned and the spider was at the entrance again. We tried the same tricks but she was having none of it. She regressed further down her burrow, fighting from within whenever I put the grass down.

It was hammering down with rain by now and we were drenched. After returning to the vehicle to get raincoats we had to make a decision. I asked Clay if we could dig the spider out in the interest of field study. There were not many opportunities to photograph these animals. He agreed after serious consideration and I took the shovel ready to carefully dig. The ground was as hard as concrete. I asked myself a lot of questions. How could a spider dig a burrow in this ground or did it take over the burrow of another spider? Maybe it made the burrow last wet season when the ground was more pliable. How long does she live in there? Months? Years? Whatever the answer, it was a wonder of nature and one to be impressed by.

Clay was a more robust chap than me and he took up the digging duties (hey – I had paid, after all) and took a few inches out of the entrance. The burrow then angled 90 degrees to the right of the main entrance and it contained young spiderlings. It was evident that this really was the time when the tarantulas were geared up for the rains. Everything was set for the young to disperse and find a good meal to start them on their way in life. A mother could do no more in that harsh, arid environment. There were plenty of predators around that would take a huge chunk out of the numbers when the young did disperse. Lizards, birds, frogs, toads and indeed other spiders would be waiting open-mouthed for a feed. Nature's strategy of safety in numbers is a way of ensuring that at least some of the young survive. It is of interest that humans often apply the same technique. Football hooligans are a prime example of safety in numbers, as are young men fighting on the street outside pubs and bars.

After some seriously hard graft the female was cornered in a side chamber of the burrow and we gently eased her out. She was a sight to behold in the dark, wet, humid evening. Light brown all over; she was neither fat nor thin. I concluded that at least something by way of a meal must have happened upon the burrow in recent weeks. She had no parasitic mites on her, and looked in very good condition. I photographed her and the burrow but was unable to determine the genus or species merely by visual means alone. I am fairly confident that it was a *Phlogiellus* sp. but it may just be *Selenocosmia*. I would have needed to take the specimen and examine its shed skin but I had no authority by way of official paperwork to take any specimens so she went back to the wild where she belonged, which was actually very satisfying. She had given up her secrets for a short time and I was indebted to her. There was no whistling, no barking, no stridulation and no aggression. We would have done a little more work regarding temperature and burrow construction but it was absolutely pouring down and we both agreed with great mutual satisfaction that we should call it a night. There was plenty of her burrow untouched for her to retire into and re-develop an entrance.

We returned to the 4x4 and Clay expressed his delight at the events of the evening. Having not been spider-hunting in such a manner before, he said that he had had a great time. He thought that had we filmed it, there would have been a short documentary in the making. We both agreed to leave the door open to possibly making this happen in the future. I may yet go back and make a short film about the life of this spider and go and catch a fish with Bill at the same time!

Clay kindly presented me with a book on Australian spiders as a parting gift, typical of the generosity and thoughtfulness of the man. I gave him a copy of my own tome about searching for tarantulas in Ecuador.

When I asked him if he did much wildlife photography he said that he enjoyed photography but when he had the incentive he never had the time and when he had the time he did not have the incentive; rather profound words that we can all relate to about something in our lives.

I went to bed that night dog-tired but relieved that the job had been done and I could now look forward to a few relaxing days in Sydney before going home.

Wednesday 19th November

We gave ourselves a day's breathing space today in preparation for the flight tomorrow to Sydney. I was very sanguine about the whole idea of getting back into an aircraft and was pleased with my coping skills.

It was a curious day – a time to reflect on what we had done and where we had been. We decided to make our last evening meal in Port Douglas a more oriental one and enjoyed a tasty curry at an Indian restaurant. (OK, I know India isn't in the Orient, but you know what I mean) Oddly enough the two waitresses were English girls just passing through town and making a few dollars on the way. One had a Yorkshire accent; not quite what I expected for an Indian restaurant but certainly novel. The reader can imagine the scenario – "Ey up luv, is it a Chicken Bhuna you want or summat else? 'Appen I'll get you t' menu lad."

We returned to the Mandalay Apartments suitably full and ready for bed but of course I was unable to let the opportunity pass to search the side roads and bushes for wildlife. It had rained profusely that morning and the delicate little bromeliads and other plants held pools of fresh water that many insects and other animals would drink from or bathe in.

I found some pretty little frogs, but was unsure what species they were. I also came face to face with a large gecko pressed motionless against the branch of a tree. Imagine if we could lick our own eyes. How impressive or repulsive would that be when trying to chat up a member of the opposite sex?

My final memories of searching amongst the undergrowth in this area were to find a variety of spider species including some I had never seen before. There were comb-footed spiders and many orb web spiders, and best of all, one single tree brimming with long-jawed spiders of the genus *Tetragnatha*. These are completely harmless to humans despite the size of their enormous jaws. I had seen many pictures in books but it's better to see them live and in their natural habitat. It helps to know where to look next time.

Chapter 9
The X-Factor Debacle

Thursday 20th November

I have already described how much I loathe flying, but it is a cursed necessity if anyone wishes to travel places quickly.

We faced a short three-hour flight to Sydney, which was some 2473km away from Cairns – don't worry, I checked. First we had to say goodbye to the management team at the apartments, to the scrub fowl that woke me each day with a delightful throaty song and to the glorious land that was Far North Queensland. I really hoped that I would return to do some filming with Clay but for now I was satisfied with all that we had done in the time available. It was time to put away my rainforest clothing and don the slightly warmer gear ready for the big city. This was to be Susan's main reason for travelling all this way – to see her very good friends Jess and Sean and the lively black labradoodle called Woof that they had recently welcomed into their home.

The flight was fuss free and there was very little turbulence, which pleased me greatly. Call me sexist if you like but it seems to me that stewardesses are much prettier than the ladies on European flights – but then even that has changed and it's mostly men nowadays. Not that I have a real issue with it but I should explain something. My fear of flying used to be so ridiculously irrational that I once figured that if the trolley-dollies were absolutely gorgeous, there was no way that any kind of merciful God would let them die in a horrible plane crash. See – that's what real fear of flying does to you. Others will give you equally ridiculous reasons. Somehow, some way you have to find a reason to believe everything will be all right. That's not to say that aircraft with crew comprising of men or plain ladies is doomed to crash. They all do a job that I wouldn't do and I respect them for it. I have no idea why anyone would want to be spilling – sorry serving coffee up in the sky but each to their own. All that turbulence, all those germs circulating around, all those screaming kids, the endless safety demonstrations and the constant requirement to look smart and slap half a whale on their faces (the ladies not the men) must get them down at times. They do have an advantage though – no waiting in airport queues for them. Anyway – they were all thoroughly nice people and I needn't have worried. We touched down in Sydney in fine fettle and I was beginning to get back into the flying rhythm after so many years. We were due to be met by a taxi driver and we were told to look out for a tall gentleman in a red hat. Such a man emerged from the throng when we exited the gate and he immediately took the lightest bag (Angie's) on the kilometre walk to where the minibus was. Meanwhile, Sue, myself and a tiny slip of a girl with mountains of heavy baggage, were forced to traipse like refugees to the stifling vehicle that just passed for a minibus.

Thanks very much for that one, mate. Thanks for not helping the smallest person with the heaviest baggage. He was either Spanish or Mexican and we seemed to have clashed with siesta time. To make matters worse we had to wait for another couple that was arriving at a different terminal so we sat outside in the heat for twenty-five minutes until the dim outlines of three human frames began to get closer. They were a couple of elderly Australians and we thought from the tone of the conversation that they and Red Hat had exchanged verbal bullets in some kind of contretemps in the building.

We drove into the middle of Sydney and on our way the driver removed his nice red hat and laid it on the seat beside him in the front. It was a kind of cowboy hat; bright red with black trim. It was a fine item of apparel but wearing it down my local high street might constitute an infringement of some local by-law and trigger a visit from the fashion police. For the next half an hour he engaged in taking consecutive mobile phone calls, which distracted him from his

The mosaic roof of the Opera House

driving and he steered us (literally) towards some pretty precarious situations.

We stopped outside a grotty looking building next to a kebab shop. It could have been a remnant of a mortar attack in a war zone for all I knew. The driver smartly placed his red hat back on his cranium and slid open the minibus door with a sudden burst of energy and enthusiasm.

"This is it," he said, nodding toward the deeply depressing construction to the left of us.

"This is your hotel," he said, looking at the Aussie couple.

"Oh," said the man, palpably shocked. "Why didn't you put the bloody cricket on the radio?" he enquired.

"This is your stop," repeated Red Hat.

I looked at Sue, who looked at Angie and we all thought the same thing. We were so glad we were not staying there. It wasn't snobbery, but a fact...it looked rough but may have been quite pleasant inside. However, we didn't want to chance it.

Besides, we hadn't seen ours yet and it may have been some kind of hell on earth for all we knew. It turned out to be reasonable indeed for a city hotel, but I will admit that we had been utterly spoiled by the Mandalay back in FNQ.

Oh well – the trick was not to spend much time in it anyway and get out and enjoy the city. An old colleague of mine used to say about hotel rooms "They're all the same with the bloody lights off," and he has a point. It was several degrees colder here, and raining too. It felt a bit like England so we were used to it. Mind you, we never had any warm clothing and had to purchase one or two items with all haste. Clothes were not too expensive as a rule but shoes were, as were food and books. This was the time of a world recession, or so the politicians kept telling us, so getting a good price for something was an extra incentive.

We unpacked hastily and headed off for the best known tourist attraction in Sydney: the Opera House. It is a unique building. I have always wanted to stand before it and admire it in my own time, in my own way. We walked briskly down the street and first saw Sydney Harbour Bridge, an imposing construction, dark and foreboding against the grey sky. It is known locally as 'The coathanger' and is the widest long-span bridge in the world at 151.3ft. I am delighted to announce that the rivets were made in Lancashire, England. Even further, the bridge itself was designed by Dorman Long & Co Ltd of Middlesbrough, England, and it opened in 1932. I am told it still carries rail, road and pedestrian traffic and of course the view of the harbour is meant to be splendid. If you were to walk it, the bridge is 1149 metres in length.

We passed the ferry port where a young Aborigine was playing his didjeridu to music emanating from large modern speakers either side of him. He was good – very good – and I admired his talent and entrepreneurial spirit.

A Daintree Diary

Finally we rounded a corner and the Opera House came into view. It was a moving moment for me, like the time I first set eyes upon Red Square in Moscow when the tears literally welled up in my eyes. We all have defining moments in our lives...what was yours? Where in the world would literally take your breath away?

Here are a few points about the Opera House that you may or may not know:

- It was built in 1973
- The architect was Danish – his name was Jørn Utzon
- It cost $102 million to build
- It is a World Heritage Site
- The roofs of the house are covered in Swedish tiles – all 1,056,006 of them
- The building covers 1.8 hectares of land.
- Its power supply is equivalent to a town of 25,000 people
- Utzon died on 29th November 2008

I photographed it like any tourist worth their salt would, but I wanted to touch it, to feel it. I wanted something more tactile, more personal. As we got nearer we saw a huge scaffold construction and a stage being erected to the side of it. There were workers busying themselves and we assumed this was all preparation for some grand opera such as *Aida* or the like. I decided to ask one of the workers and he said "We've got the Australian X-factor finals here tomorrow, mate." I thanked him for the warning and assured him I would be nowhere near the place when the unedifying spectacle took place. How dare they desecrate such a unique site in such a profoundly insulting manner? I suppose it is what the world has come to. We have gone from the halcyon days of learning how to make fire and tools and clothing houses and wheels, and food, and ale and fighting weapons to watching people live out their own lives singing absurd and nauseating crap on X-factor.

Back to the Opera House: after all this time (45 years) and all this way (10,000 miles) I laid my right hand on one of the cream tiles and closed my eyes. I listened to the sounds around me. The distant sound of a ship leaving the harbour, the chatter of fine ladies drinking wine on the harbour front, the sound of gulls crying out overhead... and the bloody clank clank clank of scaffolding going up for X-factor.

I would not allow it to diminish my love of the building or the magic of the moment. Sue, Angie and I took our obligatory photographs and agreed to walk by the botanical gardens where we could obtain photos of the Opera House from a more distant perspective. We made our way through throngs of excited young people, dressed in their best ready to go inside for an event of some kind. As we walked on, several coaches pulled up and people evidently going to the same event spilled out onto the pavement all looking terribly smart. There was a wedding and a prom going on. No one moved out of the way for us mere 'oiks', which annoyed me somewhat. One of them who happened to be an American got out of the coach that his buddy had just ejaculated from and said as loudly as possible "Hey, fancy meeting you here," whilst pausing for everyone to look at him. I couldn't resist shouting over "He was sitting behind you on the bus, you prat," which was met with visual condemnation and an upturned nose. It was great fun and I do love being a miserable old git at times.

The first sign you see at the gardens constitutes a breath of fresh air. It reads 'PLEASE WALK ON THE GRASS... We also invite you to smell the roses, hug the trees, talk to the birds, sit on the benches and picnic on the lawns.'

It was a far cry indeed from the prohibitive attitude in Britain where 'Keep off' and 'Private, keep out' signs seem to be the norm. It brought home to me how many of our signs indicate what we may not do. Let's have more that tell us what we can do. Now that would really stuff up the local councils. We could have signs like 'Please Park Here' and 'This machine does give change.' What about 'Park at our risk' or even (wait for it) 'Wheel clampers do not operate in this area.' There's more chance of teaching a monkey to use a computer than that happening.

We sat on a bench by a walled garden to 'people-watch' for a while. I became restless and looked on the wall behind the bench. There was a rather jaded spider's web with a hole about the size of a ten pence piece at the corner. My curiosity was suitably aroused as I detected several 'trip lines' radiating from it. There were also some shiny stout black legs tucked at the entrance. My heart thumped. Could this be a live specimen of the infamous Sydney funnel web spider *(Atrax robustus)* that I was desperate to see?

It was time for the old Indian trick and I grabbed a thin piece of twig and touched at the entrance. WHACK! The spider flew out and bit hard on the wood then ran across the wall in a flash. When it came to rest I photographed it for later scrutiny. Alas, it was not a funnel web but it was a fascinating, ominous-looking spider. I still do not know what it was. Again, the thought that we are seldom far away from a spider came to mind. The girls were revolting and dragged me away. I was extremely naughty and 'jogged' off impersonating a bloke that was passing, waving my arms about and stretching occasionally trying to look really cool, but failing. The ladies were in fits – I still have the gift for comedy after all! The urge to eat saw us head off in the general direction of the hotel for further sustenance.

We dressed for dinner and visited a Chinese restaurant nearby. It might appear that we never ate 'typically Australian food' whilst on our visit but we certainly did. You have to ask yourself what typically Australian food is. It is much like ours in Britain – barbeque meat, fish and

A Daintree Diary

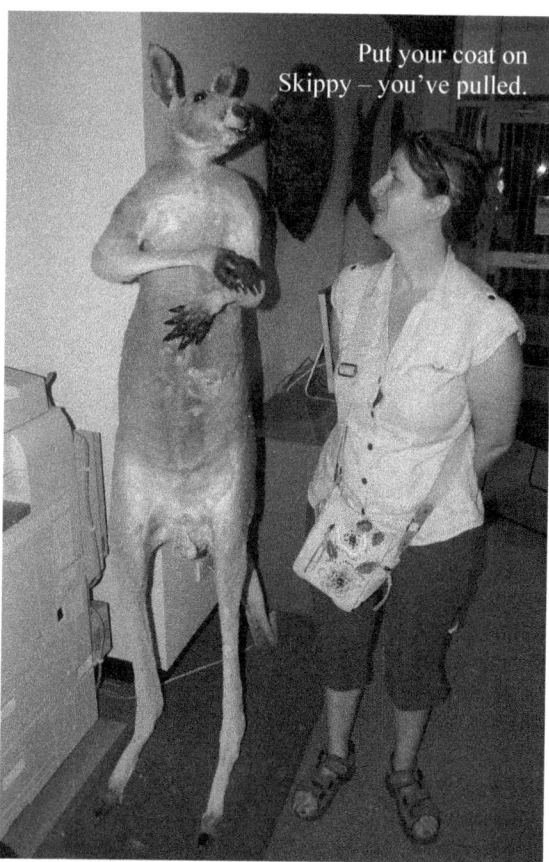

Put your coat on Skippy – you've pulled.

prawns etc – just fresher and tastier. Sue is pescetarian so meat was out of the question. Angie was a red-meat-eater like me, so that was not a problem. I got used to her line at the end of a meal that she particularly enjoyed. "That was bloody lovely," she would exclaim. The meal was excellent and the waiter, who I think was the manager, took a shine to Angie, saying how young she looked etc. He complemented Susan too but like a typical male I thought – 'get on with it Casanova and bring the spring rolls out.'

Of course, he might have pulled if he had been able to lick his own eyes. Another fine and invigorating day had been enjoyed by us all. The next couple of days would see Angie doing her own thing in Sydney whilst Sue and I would catch the ferry to Manley to see her friends. Before that I wanted to visit the Australian Museum and Sue was keen to come too so it was with these thoughts that our heads hit the pillows and our bodies shut down for well deserved sleep.

Friday 21st November

I love museums; particularly those with large natural history exhibits. Sue and I had decided to dedicate the whole morning to visiting the Australian Museum before catching the ferry to see Jess and Sean. Walking through the entrance I was rather taken aback to come face to face with an enormous white-bellied sea eagle. Sadly it had had a previous encounter with a taxidermist but it was impressive even in its immutable state. At least I had the pleasure of observing those live specimens in previous weeks, which was infinitely more preferable.

I took the opportunity to take photographs of the spider and insect collections, which gave me a better idea of the variety, size and habitat of the creatures that inhabited the continent. I was rather taken with a specimen of a bent-wing swift moth *(Zelotypia stacyi)* and a handsome netcasting spider of the genus *Deinopis*, which I was very surprised to learn can span the palm of an adult hand. I saw a real funnel web at last but it was in its burrow and I could only see the side of the body pressed up against the glass housing, though it gave me useful information about burrow size and construction, etc.

Hands up if you know who Frank Hurley is. No? Neither did I until we were lucky enough to visit the museum just at the time when an exhibition of his was on. James Frances Hurley (1885-1962) was one of the greatest Australian photographers of the twentieth century. His coverage of the Mawson (1911-1913) and Shackleton (1914-1916) Antarctic expeditions and the horrors of the First World War have guaranteed him his place in history. This genius was born in Sydney in 1885 and demonstrated his adventurous spirit by running away from home at the ripe old age of thirteen. He took up photography and travelled widely.

The display we saw was called 'Journeys into Papua', which in photographs showed the culture clash between the Papuan people and the colonial settlers. One could experience the historical narrative of Papua in the 1920s accompanied by 80 black and white photographs, many of which have never before appeared on public display. There was also some extremely rare film footage showing 'natives' in canoes paddling towards the camera, spears at the ready.

What a man Hurley was! Imagine his feelings at the time. He could have been killed and even eaten but he had to get that footage; he simply had to focus (literally) on the photographs. And there were Sue and I in the year 2008 standing in front of those images. Goosebumps appeared on my arms as I really thought about that. There were inspiring photographs of tribal people with decadent headgear and shrivelled skulls hanging by their huts. It was a stunning display and literally stopped me in my tracks, which is no mean feat.

While Hurley's superb coverage of the Antarctic expeditions and the tragic horror of the First World War have guaranteed him a special place in the history of photography, his work in colonial Papua is less well known. I was so impressed by this man and his work that upon my

Tell me your secrets you magnificent bird.

return home I spent a considerable amount of money on the rare book *Pearls and Savages*.

I am often embarrassed by my lack of knowledge of many things, and in some ways apart from myself I blame the media. When one thinks of 'famous' Australians most people on the street would think of Rolf Harris, Kylie Minogue, Steve Irwin, Crocodile Dundee, Russell Crowe, Hugh Jackman and the like. However, people like Frank Hurley are not even known in most quarters yet he is the biggest star of them all in my eyes. Then there was Ned Kelly, the bushranger who goes down in Australian folklore as both hero and villain. Born in 1854 and hanged in 1880 he defied the colonial authorities and I am led to believe that his last words were "Such is life."

Alas, we simply ran out of time again and managed a brief sortie to the museum shop where Sue was good enough to purchase my Christmas present. It was a resin paperweight with a male Sydney funnel web (obviously dead) inside. It now sits on my desk reminding me permanently of Australia. On the way to the exit we happened upon a very large stuffed male kangaroo rearing up on its hind legs. It stood taller than Sue and I liked its pose so I asked Sue to stand next to it for a photo. Secretly I was imagining using this for fun at a later date with captions like 'Sue and her new boyfriend.' Suddenly my eyes were drawn to something hanging from the kangaroo's body – between its legs. It wasn't a pouch that's for sure. It was a male all right. Sue had her photo taken next to a roo with a monstrous brace of onion bags. I was not only highly amused but also insanely jealous.

Page 170: And it's goodnight from me...
Page 171: Woof – as inquisitive as ever.

There was one other brief moment of comedy. Amongst the huge photographs hanging on the walls near the exit was one of a small aborigine with spectacles and a cheery smile. Susan proffered the thought that it was Ronnie Corbett in disguise and it truly did look like the twin of the great (or little) man.

The warmth of the sun enveloped our bodies once more as we stepped out of the doors, blinking in the light. It was time to catch the ferry across to Manley Island where we would be staying overnight. Whilst on the ferry it is possible to view the Opera House from new and interesting angles then sit back to enjoy the view.

Half an hour later we docked in Manley and this was the part of the trip that Sue had really anticipated. I had met Jess and Sean once in England at a party and I knew that they were a first class people. Both young (damn them); Jess is an attractive Australian with bright eyes and a winning smile, whilst Sean is a New Zealander with a mop of naturally curly hair, a great physique and an enviably relaxed attitude to life. There is nothing that these two can't or won't do to make you feel at home. The only time they seem to disagree is when sporting events such as rugby or cricket see the Aussies and their close neighbours in mortal combat. That's fair dinkum, as they would say.

They married in 2003 and I am delighted to report that a third member of the family is on the way. There is also a family member with four legs and a thick black coat. Woof is a stunning black labradoodle with a lovely temperament and the energy of a shooting star.

For those that don't know a labradoodle is literally a cross between a Labrador and a poodle and the wonderful thing is, they don't shed fur. Seeing him evoked thoughts of our own quadruped, Darwin. He is a young Border Collie with a zest for life and a nose for trouble but we missed him terribly.

If there is one thing the Aussies, the Brits and the New Zealanders have in common, it's the love of a good barbeque. The experience is made more memorable when you have a 'Barbie Master.' I am not talking about the make of the machine on which the food would cook; I am talking about Sean who effortlessly manages to cook prawns and various types of meat, whilst pulling beers and entertaining with his hilarious anecdotes.

We sat around the table chomping our way through a fine feast whilst Woof zig-zagged his way around the room with his football.

The right moment came for me to excuse myself, leaving the three of them to catch up on news whilst I went for a walk around the garden to see what arachnids I could find.

Sean told me he had seen some spiders with red abdomens lurking behind one of the fence posts and I was eager to locate them. I looked hard and found a couple of interesting garden

spiders but no 'reddies.' Sean came out and took me to the post. I pulled the plank back to reveal the red devils...but they weren't spiders.

They were assassin bugs and I was surprised to find them in such a location. I used to breed these back home and marvel at their strategy for taking large prey. They work as a team to immobilise the prey and basically tear it apart. I have a very healthy respect for these single-minded insects and I gently returned the plank to it original position and left them alone. I told Sean not to worry (as if he would!) and he went back to the girls as I carried on poking around in the grass and under stones – not necessarily too clever in Sydney funnel web country, especially at night.

Alas, I found no funnel web but I did find two extraordinarily beautiful garden spiders. One had a huge abdomen and was, I think, some kind of curious 'egg' spider and the other was the most beautiful garden spider of the genus *Araneus* that I have ever seen. It had chocolate-brown and beige legs, and a toffee-coloured body. It was covered in fur and looked like a shaggy spider. It sat perfectly for 'daddy' as we shared a Nikon moment.

Page 172: A splendid garden spider
Page 173 top: An arachnid found in Sean and Jess's garden
Page 173 bottom: This spider was busy constructing her evening web

Page 174 top: The Three Sisters
Page 174 bottom: Sean and Jess – a perfect match.
Page 175: A wonderful vista of the Blue Mountains

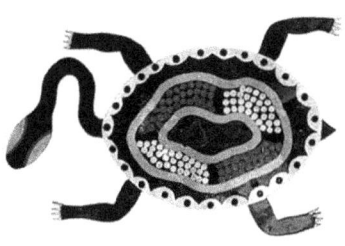

Chapter 10
The Blue Mountains

Saturday 22nd November

We had spent a very hot night in bed anticipating our trip to the famous Blue Mountains. First we were to enjoy a hearty breakfast and a bracing walk along the beach with Woof. Our hosts showed a remarkable aptitude for making toast at just the right texture, and with great speed too. We drove a few miles to a communal park adjacent to a small bay where quaint boats of many colours were moored. Jess informed us that this was a very upmarket area but no one cared that people walked their pooches on the grassed area and the dogs had an absolute ball. It was already very warm but the sky was bruised, which did not bode well for a trip to the mountains. It began to rain a little but soon stopped and began to clear. It was comical: as soon as the first drops fell most of the Aussies ran for the sanctuary of their cars whilst Sue and I flew the flag for England and stood there. We are both from the Midlands and not going to be beaten by a spot of rain. It could be argued that we were the only fools, but that's another matter! Jess pronounced that if it were raining at the Blue Mountains it would not be worth going as we literally would not be able to see for more than just a few metres.

The Blue Mountains are majestic sculptured cliffs that lie some sixty-five kilometres west of Sydney and 200,000 hectares of it are reserved as National Park. There are ridge-top and woodland walks, rock pools and rainforest. I was greatly interested to learn why the Blue Mountains are so called. It certainly isn't because they are blue (they're not). It is because the fine droplets of oil produced by the eucalyptus trees have a blue tint. The setting sun ignites the colour even more.

Jess had warned us that it would be very cold, but I did not dress for anything other than the hot temperatures we had experienced in the far north. This was a serious error of judgement. During the drive there Sean reminded us how cold it would be, and there was rain too but I just said I was English and therefore used to rain and cold, laughing off the idea of having to wear a fleece.

We drove to the main lookout point and stepped briskly out of the car. Oops! In a few seconds I was shivering but trying to keep a brave face. I concentrated on my trusty camera and getting some great shots from the lookout point. The first thing my eyes pulled focus on was the jagged vista of the Three Sisters. These are literally three jagged rock constructions, reaching for the sky and beyond them is a deep valley leading to Katoomba.

Moving around the vantage point one can view forest down below stretching as far as the eye can see. It looks like the Grand Canyon. I would have loved six months in the thick of it studying the spotted-tailed quolls, platypus, golden whistlers, water dragons and other animals that reside there. There are also koalas, kangaroos and sugar gliders in the more open woodland and forested areas. Whilst there are no seasons in Far North Queensland, the Blue Mountains are a very different matter. Here the gardens and parks are painted in frost and the leaves turn red and gold. Looking from the comfort of my lofty position I visualised how that would look. Surely there could be no better place to connect with nature and contemplate life, the universe and why no one ever sees baby pigeons.

The Blue Mountains are home to the world's steepest train and it travels almost vertically – I kid you not. The train itself looks in design like a roller coaster with little 'buckets' to sit in attached by a system of pulleys and counterweights. You begin the short ride horizontally before surging over a precipice and down into a tunnel through the rock. Sean and I loved it but I am not sure the ladies did. The ride down to the forest takes just a few minutes but is very memorable. We stopped off for a forest walk and Sean and I looked for the elusive funnel web spider. Not the Sydney funnel web but the Blue Mountain funnel web *(Hadronyche versutus)*. This is often mistaken for the Sydney funnel web but *H. versutus* has a highly arched head region and there are other taxonomic differences too. Alas, no spiders were found but there were hundreds of curious holes about two centimetres in diameter in the banks by the roadside.

Like all things tremendously enjoyable, it was with sadness that it had to come to an end. We were back in the car and travelling back to the ferry to say *au revoir*. There were two more brief stops before we got there, though. The first was a lovely surprise: Jess took us to her 'special place', one of her favourite childhood haunts. After traversing winding roads and secret groves we arrived at a cliff top covered by trees with a protective wall at the top. The view was mind-boggling as we looked across a truly vast area of forest. The trees were burnished brown and red, and the hills and mountains in the distance framed the picture perfectly. We were indebted to them both for going out of their way to bring us to such a personal place.

The second location we went to was a very surreal hotel that seemed to be influenced by designs from different countries around the world. One could observe the designs of an Indian Temple here and a typically grim 1960s British building there. The structure had a little of the *Amityville Horror* about it. The dining room was almost empty apart from one family dressed in black huddled around a huge tiered tray loaded with the most wonderfully coloured iced cakes. It was the sort of thing you would expect Alice to eat at the Mad Hatter's Tea Party.

God's waiting room

The staff members were also dressed in black and I swear every one of them had dark rings around jaundiced eyes. No one actually looked us directly in the eye. Strange 'old time' jazz music was being piped through cobweb-covered speakers.

I joked with Sean that possibly we had all been involved in a car crash on the way down and been killed only to find that this is what happens to you afterwards; a sort of waiting room to meet your maker. Our day of reckoning was at hand and we awaited our fate. This place was truly beyond the realms of death.

Sean laughed this off and commented that he thought he had seen the manager of the New Zealand rugby team only to be informed by Jess that they were actually in Wales at the time. Sean was having none of it and refused point blank to let the idea go. I could stand the uncertainty no longer and approached the man myself and said "This is going to sound rather obscure, but are you the manager of the New Zealand rugby team?" He looked at the lady opposite, who I assume was his wife, smiled and said "no mate, sorry". I got the feeling he had been mistaken for the rugger chap many times but seemed, anyway, to enjoy it. Sean was visibly destroyed to learn that he was not in the presence of his hero and I felt like a scoundrel for shattering his dream. I hate giving good people bad news.

The spookiness did not end there. Remembering that earlier in the day we had walked in warm weather on the beach, we stepped out of the hotel to be greeted by a snowstorm! It was completely surreal and I couldn't help but recall that scene in *Poltergeist* where the haunted house, built on an Indian burial ground suddenly imploded on itself in the midst of a terrifying storm.

At the ferry departure point at Manley Wharf we bade sad farewells and agreed that it had been tremendous fun. When we got to the other side we bought coffee and paninis and sat in the (now sunny) park and took in the sights and sounds of the city. It was a lovely but brief pause before the long journey back to London.

Later that evening, upon our return, Angie and I watched the rugby league world cup final between New Zealand and Australia. The match began with Australia walking directly within a metre of the Kiwis when they were performing the Hakka. Unfortunately this act of defiance did little to put the opposition off and New Zealand won the match.

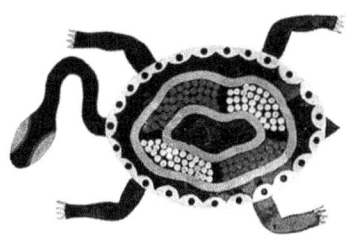

Chapter 11
Return to Londinium

Sunday 23rd November

The weather was dire today and it was actually a good day to leave Sydney from that perspective. I felt more comfortable and more confident about flying by now and I accepted that it was just something that had to be done to get from A to B. The plan was to fly from Sydney to Bangkok airport where we would have only an hour for refuelling the aircraft, and journey on to London Heathrow.

A very funny event happened as we checked our baggage in at Sydney airport. Angie's case was heavy – she suspected too heavy but had been fortunate to get away with it a couple of times except one. As we approached the baggage check-in there was a man inspecting every single case with weighing scales.

He was either Chinese or Japanese and Angie said "Ooh, I don't like this…look at him checking every bloody bag." She shuffled nervously and failed to hide the fact that she was on edge about this development. Just before we got to the official the airline opened another line and we darted to it to escape the rigorous inspection. The man did seem to be insanely protective about every single kilogram and he was as dedicated to his job as a parasite on the skin of its reluctant host. Angie let out a big sigh of relief and as we went through she glowered at the man and said "Does he want his cheque back?"

"Sorry?" I enquired…

"From charm school" snorted Angie in her annoyance.

I laughed aloud. Angie can be at her best in moments such as this.

On the flight to Thailand I started to think about the prospect of having to get back home, back to work and back to the real world. I tried to think of the words of Henry Walter Bates in his book *The Naturalist on the River Amazons* where he described his own feelings about returning to England after some eleven years in South America in the mid-nineteenth century. I am able to quote directly the words he said:

"Pictures of startling clearness rose up of the gloomy winters, the long grey twilights, murky atmosphere, elongated shadows, chilly springs, and sloppy summers; of factory chimneys, and crowds of grimy operatives, rung to work in early morning by factory bells; of union workhouses, confined rooms, artificial cares, and slavish conventionalities."

I really did not want to return and could have cheerfully spent many more weeks and months in Australia.

We proceeded to fly into Thailand and were pleased to be told that we could disembark for a short time and have a walk around the few shops at the airport. Our first observation was significant numbers of people lying on the floor sleeping or obviously preparing for a long stay; not like they were awaiting a flight, but actually living there. I do not know what religion or nationality the people were but they could well have been Muslim/Sikh/Hindu families with children and elder people sleeping on makeshift mats. We thought this was curious at the time but didn't make any connection to what was about to happen at the airport.

We flew out and were soon on our way – evidently one of the last flights to do so for many days because shortly afterwards the airport was taken over by dissident factors who were angry with the Thai government. No one was allowed in or out of the besieged airport for well over a week and it was headline news all over the world.

We were lucky but many holidaymakers were completely stranded and running out of money fast. It just goes to show that one must be prepared for the unexpected when travelling abroad and knowing a little of current affairs via the Foreign and Commonwealth website is never wasted effort.

Monday 24th November

As anyone who has endured it is aware, it's a long old flight to get home and all you want to do is be back in your own bed. It seemed to go on forever and does not help if the films are pretty average. I listened to my music on my MP3 player. For the record (no pun intended) I had *Saxon, Judas Priest, Metallica* and *Benedictum* as my heavy metal songs and Smetana (Ma Vlast) Mozart's Requiem in D minor and Tchaikovsky (Marche Slav) for afters. What an eclectic taste I have.

We disembarked to be met by an evil Baltic wind and driving icy rain. Had it even changed during the month we were away? Our taxi was delayed and we had to wait in a freezing semi-open enclosure, exposed to the weather. I said to the girls "Welcome home. How long before we are greeted back into our own country?"

With that a young man waltzed around the corner with his suitcase, obviously in a hurry. "This fucking place, where's the terminal?" he said. "Ah; welcome back to Britain, ladies," I added, devilishly satisfied that it had taken but minutes to hear our first F-word back in Blighty.

"You what?" he replied.

I said "Don't worry about it, mate. I just wondered how long before we heard the F-word and you didn't let me down. The terminal is down those stairs."

The Neanderthal glowered and disappeared out of our day and indeed out of the rest of our lives, which gave cause for celebration.

There are many species of scorpion is Australia, yet none are really considered particularly dangerous to humans. They have been around, largely unchanged for millions of years and they are great survivors.

I love one of Aesop's fables about the scorpion which goes like this:

A scorpion asks a frog to carry him across a river. The frog is afraid of being stung, but the scorpion reassures him that if it stung the frog, the frog would sink and the scorpion would drown as well.

The frog then agrees; nevertheless, in mid-river, the scorpion stings him, dooming the two of them. When asked why, the scorpion explains, "I'm a scorpion; it's my nature."

Chapter 12
Post Trip Comments

I am smitten with Australia. For a brief moment in the maelstrom of time I experienced some of the magic that the country has to offer. People have argued that it is a largely uninhabitable and hostile land with brusque and arrogant people. The old joke about the Aussies descending from criminals is actually wearing thin but they even joke about it themselves. The people I met were generous to a fault and went out of their way to help. They helped me to fulfill another dream and I could not have written this account without them. They love life; they enjoy their sport, food and humour the same as anyone else. They respect their precious environment and they are knowledgeable about the land and what it has to offer. I admire that.

Just a few short weeks after we returned I came across a poem by the British Poet Laureate Sir John Betjeman, which seemed to encapsulate the very essence of my return trip.

Back from Australia
Sir John Betjeman

Cocooned in Time, at this inhuman height,
The packaged food tastes neutrally of clay,
We never seem to catch the running day
But travel on in everlasting night
With all the chic accoutrements of flight:
Lotions and essences in neat array
And yet another plastic cup and tray.
"Thank you so much. Oh no, I'm quite all right".
At home in Cornwall hurrying autumn skies
Leave Bray Hill barren, Stepper jutting bare,
And hold the moon above the sea-wet sand.
The very last of late September dies
In frosty silence and the hills declare
How vast the sky is, looked at from the land.

In many ways, Australia is a young country with 'no real history' (to quote an Australian) but it has the potential to have the most incredible future. It is down to the will and foresight of the people and I for one wish them nothing but good luck.

I hope with all my heart that the Queensland rainforest is protected by the Australian people for the generations to come. It is a most horrifying thought to me that yet more species of plants and animals could become extinct and that people like me, a back-street city boy from the other side of the world, would not have the opportunity to set foot upon its breathtaking landscape and experience one of the true wonders of the world.

I have always intended to end this book with words of wisdom. They are not mine, but belong to the old Cree Indians, who must have the final say:

> *Only after the last tree has been cut down*
> *Only after the last river has been poisoned*
> *Only after the last fish has been caught*
> *Only then will you find that money*
> *Cannot be eaten*

The End

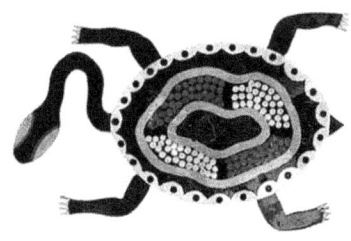

ABOUT THE AUTHOR

Carl Portman is the author of *Fangs for the Memories: The Search For Tarantulas In Ecuador*. He has bred many species of tarantula and scorpion and gives lectures on arachnids in schools and other institutions. He has helped with programmes such as *Giants* on British television and has also worked with TV and radio in Germany, illustrating the positive aspects of spiders and their kin. He is a former committee member of the British Tarantula Society and regularly contributes to the society journal.

He lives in Oxfordshire with Susan; 'Darwin', their Border collie; and several tarantulas. When not out photographing nature he can be found burning the midnight oil writing or playing internet chess.

His other addiction, from which there is no cure, is Aston Villa Football Club and can regularly be found at the famous Holte End cheering on the boys in claret and blue.

His motto is '*Get up, get dressed and get out there.*'

Contact: carl.portman@hotmail.co.uk

Also by Carl Portman...
**copies available direct from the author, email
carl.portman@hotmail.co.uk**

Paperback: 104 pages
Publisher: Square One Publications (26 Oct 1998)
Language English
ISBN-10: 189995533X
ISBN-13: 978-1899955336

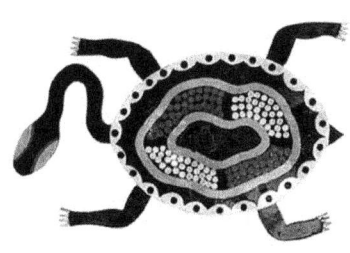

REFERENCES

Flying Without Fear - Captain Keith Godfrey
A field Guide to Singapore Zoo - Singapore Zoo
Wildlife of Tropical North Queensland - (A Queensland publication)
All about Didjeridus - Boongar
Daintree, Jewel of Tropical North Queensland
Collected Poems - John Betjeman
The Naturalist on the River Amazons - Henry Walter Bates
Fangs for the memories - the search for tarantulas in Ecuador - Carl Portman

www.wikipedia.org
www.cairns.com.au for the story of the nephila
www.kurandascenicrailway.com.au for the railway information
www.kuranda.org for the Kuranda and Skyrail details
www.thebts.co.uk for the British Tarantula Society

I love this with all my heart. If you could have seen the expression on my face when first I saw it you would have seen just how awestruck I was. It encapsulates the whole trip for me.

If there had to be a title it would be called 'It's not enough to look, you have to see - so what DO you see?'

Carl

APPENDIX: Using the GETAC B300 in the rainforest

GETAC recommends Windows Vista® Business.

"Recommended"

It became a trusty friend and I would highly recommend it to anyone who needs a robust and rugged laptop for fieldwork. The night vision capability was invaluable and saved a lot of time and rework back at base.

Case Study – Carl Portman

Sand, sea and snakes
The GETAC B300 goes down under

Background

As a lover of natural history, and entomology in particular I take as many opportunities as I can to see animals and plants in their natural environment. I am fascinated and intrigued by rainforest environments and I had a burning desire to visit the oldest one on the planet. The Queensland forests have existed for some 120 million years. When Australia broke free from Antarctica about 50 million years ago it carried with it a cargo of living plants and animals. It is the only rainforest left in Australia and represents only 0.01% of the total land mass – so it is incredibly important.

This is a place where rainforest meets the sea. It is a land of strange creatures such as the duck billed platypus, the cassowary and the mudskipper and contains some of the oldest plants on earth such as cycads which are survivors from millions of years.

The Daintree area is the jewel of tropical north Queensland and much of it is inaccessible wilderness. It was named in honour of an Englishman, Richard Daintree who never actually saw it!

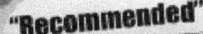

BUILT TO SURVIVE

www.getac.com

GETAC recommends Windows Vista® Business.

The environment is hot and humid and has one of the wettest climates in Australia. I went in November however, just before the wet season and the temperature was always between 28˚ and 33˚ C.

I wanted to take a laptop computer with me to enable me to input work including photographs – in the field as they happened. Any old computer would not do and would fall apart in the unforgiving environment (the rainforest is a war zone for the flora and fauna it contains) that I would be in. If you stand still for too long here something will grow on you.

GETAC kindly accepted my request to loan a laptop and try it out in rainforest conditions. Here are my thoughts then on the B300. This PC is built for rugged environments and usage. It had its first test at Heathrow airport when the rollers going through the x-ray machine decided to reverse and spew it back out and onto the floor! Okay, it was in a padded bag but nevertheless other laptops may have called it a day there and then. Fortunately there was no problem for the B300. It was interesting to note, upon my arrival at Darwin airport that everyone who had laptops had to take them out of cases before they were processed through the x-ray machine. The young lady operating the machine saw the GETAC, picked it up and said 'hey, that's a very nice laptop' before letting it go. I felt superior for a fleeting moment as I looked around at the other 'basic' models.

[Figure 1] Tropical Rainforest at Kuranda, Queensland

What environment did I use it in?

The answer is anywhere I could. This included the beach in the searing hot sun. I left it in the sand and washed it in the sea. It was in the rainforest by day and night in hot humid conditions and occasional rain. It was left on rainforest floor leaf litter and sand and soil as well as getting wet on one particular occasion when the heavens opened for only a short while.

[Figure 2] Waterproof and dust resistance

Rugged Mobile Computing Solutions
BUILT TO SURVIVE

www.getac.com

GETAC recommends Windows Vista® Business.

What was good about it?

I actually like the physical look of the laptop first and foremost. It is fairly heavy at 3.5kgs but needs to be to withstand some of the rigours of its life. I especially like the very neat night vision capability where the keyboard glows red to enable the use to continue working even in total darkness. This proved a useful resource when hunting for the Amethystine python.

It loaded quickly and has other useful functions such as eco standby and many special function keys. All the sockets and ports are covered and the CD tray is well protected too.

Built from a magnesium alloy case I had every confidence that it would be up to the job in a tough environment.

It has a long battery life, reportedly up to 12 hours but I did not fully test this on this trip. It went through many potentially harmful x-ray machines on this trip with no adverse affects to any software or documents. Finally, although I never use it the laptop has fingerprint recognition which I have never seen before.

What didn't I like?

I feel I should balance my review by pointing out a couple of things I did not like about the laptop. Firstly, the handle is moveable and pushes in and out of the pc. It is easy to nip your skin between the handle and the body of the laptop, and that's exactly what I did, very painfully on one occasion.

The second point is that the USB port on the right hand side is too low. When you open the sealed port rubber casing and insert (for example) a memory stick it is pushed upwards by the outer flap and can bend the memory stick/other appliance which is not good. I do have to query how long these protective rubber covers will last as the one I used seemed a bit ropey to me. Once they have fallen away the whole unit ceases to be as penetration proof as it was.

These are only small points but of the three the most annoying was the port location. As I type – the PC is resting on a book for elevation to ensure the memory stick does not bend and that cannot be right.

I have to say that I believe the B300 could have stood a lot more but since it was loaned I did not want to push it.

A REPORT FROM THE FIELD – THE PYTHON AND THE FROG!

I spent an evening looking for the elusive Amethystine python (*Morelia amethistina*). This beautiful animal is not potentially dangerous as it does not have venom, but it suffocates its prey which includes mice, frogs, birds and even domestic chickens. They will readily bite however and do have a sharp row of teeth. With GETAC in hand I set off to find one. (figure 3)

[Figure 3]

Rugged Mobile Computing Solutions
BUILT TO SURVIVE

www.getac.com

You have to be careful in a tropical rainforest at night and always expect the unexpected. I could come across the wrong snake – a death adder or several kinds of tree snake. I could even come across a huge wild pig and they are not easy to deal with if they choose to come at you. I saw something move, but it was not my snake. It was a beautiful mature male huntsman spider (figure 4) which was out looking for a mate. This guy spanned the palm of an adult hand.

One can spend days in the forest and not find a snake but luckily within an hour there was one sitting right in front of me. Not only that but it was the species I was hoping to find. A freshly shed python (figure 5) stayed quite still as I approached and it seemed quite calm so I was pleased to be able to handle it calmly (Figure 6) and without getting it too excited as they are usually prone to biting. I asked my companion to hold the snakes head whilst I got a photo and thanks to having the GETAC I was soon able to download the picture using the red night-site keyboard. This is the first time I have used a laptop in the field (figure 7) and it was actually rather useful to input photos and other data in real time instead of having to write it all down to regurgitate later. I also used this neat facility on the flight back when all the cabin lights were off – a nice touch from GETAC. There were a few enquiring admiring glances from the passengers in the seats surrounding me.

I ventured deeper into the forest and saw a huge flying fox fruit bat with a one metre wingspan in front of me. There were many other amazing animals and plants too, such as giant cluster fig trees with huge buttress roots, and also mangroves and glorious ancient ferns.

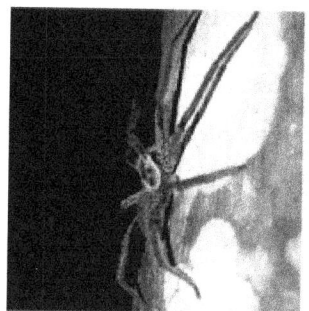

[Figure 4]

[Figure 5]

[Figure 6]

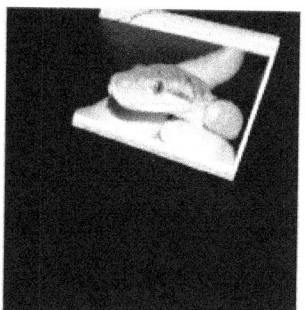

[Figure 7]

www.getac.com

Rugged Mobile Computing Solutions
BUILT TO SURVIVE

GETAC recommends Windows Vista® Business.

It was whilst I was inspecting such a fern that I turned around to find a large white-lipped tree frog (Litoria infrafrenata) had taken a liking to the B300 (figure 8) and rested on the keyboard. I had turned the night light off whilst I went away so it must have been attracted by the screen or the fact that something 'new' was in its environment.

These creatures are not often seen so it was a nice surprise but I gently took it off and set it aside ready to work. Sure enough, with one giant leap, the little star landed on the screen (figure 9) and seemed very happy there. I just ignored it and scribbled some notes in my pocket book. I then continued to work although it was terribly hot by now (figure 10) and I was about ready to go when he popped up again behind the screen (figure 11) and proceeded to hold on as if to say 'this is mine mate'. So – even the rainforest frogs of tropical Queensland are impressed by the B300 which is praise indeed.

Finally, even with the best of intentions I could not carry the GETAC around with me all the time. The weight of that plus camera equipment and bottles of water is too much on a boiling night. However I was perfectly happy to leave it at a marked place (I was in the middle of nowhere after all) and return to it on my way back. With a normal laptop I would have worried about rain or soil getting into the machine

[Figure 8]

[Figure 9]

[Figure 10]

[Figure 11]

www.getac.com

Rugged Mobile Computing Solutions
BUILT TO SURVIVE

GETAC recommends Windows Vista® Business.

and also the possibility of ants or other small animals getting in and possibly causing some damage. Not so with the B300 thankfully and I returned to it knowing that something had been mooching around it (figure 12) but that it was perfectly safe.

My thanks to GETAC for having the courage to loan this excellent machine to a complete stranger in the name of doing something different!

[Figure 12]

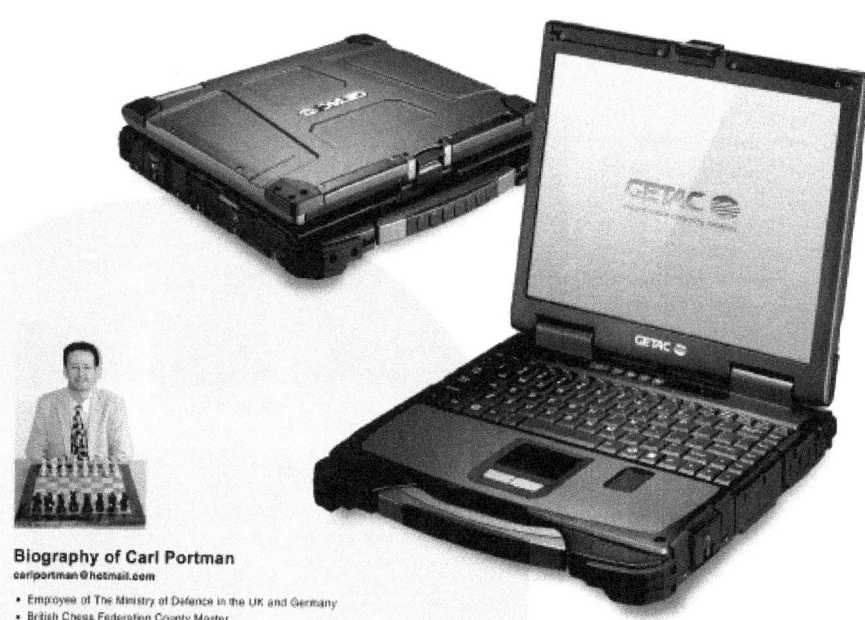

Biography of Carl Portman
carlportman@hotmail.com

- Employee of The Ministry of Defence in the UK and Germany
- British Chess Federation County Master
- Member of the British Tarantula Society
- Hobbies: Chess and arachnology. Has bred rare and endangered species.
- Author of the book 'Fangs for the memories, the search for tarantulas in Ecuador'
- Motto in life is 'get up, get dressed and get out there'

GETAC US
22762 Drive Lane,
Lake Forest, CA 92630, USA
TEL: +1-949-699-2888
Toll Free: +1-866-GO-GETAC
(1-866-464-3822)
http://www.getac.com

GETAC EUROPE
Nedge Hill
Telford TF3 3AH, UK
TEL: +44 1952 207 221

GETAC ASIA-PACIFIC
5F, No. 75,
Ming-Shang East Rd., Sec. 3,
Taipei, Taiwan, R.O.C.
TEL: +886-2-2501-8231

SALES CONTACT
- North America:
  Ruggedsales@getac.com
- South America:
  SouthAmericasales@getac.com
- EMEA:
  EMEAsales@getac.com
- APEC:
  APECsales@getac.com
- OEM/ODM:
  ODM.sales@getac.com

SERVICE CONTACT
- Global Service:
  Support@getac.com
- US Toll Free:
  +1-866-EZGETAC
  (1-866-394-3822)

MARKETING CONTACT
- Global Marketing
  Globalmarketing@getac.com
- US Marketing
  USmarketing@getac.com

© 2009 GETAC Inc. All rights reserved. V1.M01109

THE CENTRE FOR FORTEAN ZOOLOGY

So, what is the Centre for Fortean Zoology?

We are a non profit-making organisation founded in 1992 with the aim of being a clearing house for information, and coordinating research into mystery animals around the world. We also study out of place animals, rare and aberrant animal behaviour, and Zooform Phenomena; little-understood "things" that appear to be animals, but which are in fact nothing of the sort, and not even alive (at least in the way we understand the term).

Why should I join the Centre for Fortean Zoology?

Not only are we the biggest organisation of our type in the world, but - or so we like to think - we are the best. We are certainly the only truly global Cryptozoological research organisation, and we carry out our investigations using a strictly scientific set of guidelines. We are expanding all the time and looking to recruit new members to help us in our research into mysterious animals and strange creatures across the globe. Why should you join us? Because, if you are genuinely interested in trying to solve the last great mysteries of Mother Nature, there is nobody better than us with whom to do it.

What do I get if I join the Centre for Fortean Zoology?

For £12 a year, you get a four-issue subscription to our journal *Animals & Men*. Each issue contains 60 pages packed with news, articles, letters, research papers, field reports, and even a gossip column! The magazine is A5 in format with a full colour cover. You also have access to one of the world's largest collections of resource material dealing with cryptozoology and allied disciplines, and people from the CFZ membership regularly take part in fieldwork and expeditions around the world.

How is the Centre for Fortean Zoology organised?

The CFZ is managed by a three-man board of trustees, with a non-profit making trust registered with HM Government Stamp Office. The board of trustees is supported by a Permanent Directorate of full and part-time staff, and advised by a Consultancy Board of specialists - many of whom are world-renowned experts in their particular field. We have regional representatives across the UK, the USA, and many other parts of the world, and are affiliated with other organisations whose aims and protocols mirror our own.

I am new to the subject, and although I am interested I have little practical knowledge. I don't want to feel out of my depth. What should I do?

Don't worry. We were *all* beginners once. You'll find that the people at the CFZ are friendly and approachable. We have a thriving forum on the website which is the hub of an ever-growing electronic community. You will soon find your feet. Many members of the CFZ Permanent Directorate started off as ordinary members, and now work full-time chasing monsters around the world.

I have an idea for a project which isn't on your website. What do I do?

Write to us, e-mail us, or telephone us. The list of future projects on the website is not exhaustive. If you have a good idea for an investigation, please tell us. We may well be able to help.

How do I go on an expedition?

We are always looking for volunteers to join us. If you see a project that interests you, do not hesitate to get in touch with us. Under certain circumstances we can help provide funding for your trip. If you look on the future projects section of the website, you can see some of the projects that we have pencilled in for the next few years.

In 2003 and 2004 we sent three-man expeditions to Sumatra looking for Orang-Pendek - a semi legendary bipedal ape. The same three went to Mongolia in 2005. All three members started off merely subscribers to the CFZ magazine.

Next time it could be you!

Project Kerinci, Sumatra - 2003
In search of the bipedal ape Orang Pendek

How is the Centre for Fortean Zoology funded?

We have no magic sources of income. All our funds come from donations, membership fees, works that we do for TV, radio or magazines, and sales of our publications and merchandise. We are always looking for corporate sponsorship, and other sources of revenue. If you have any ideas for fund-raising please let us know. However, unlike other cryptozoological organisations in the past, we do not live in an intellectual ivory tower. We are not afraid to get our hands dirty, and furthermore we are not one of those organisations where the membership have to raise money so that a privileged few can go on expensive foreign trips. Our research teams, both in the UK and abroad, consist of a mixture of experienced and inexperienced personnel. We are truly a community, and work on the premise that the benefits of CFZ membership are open to all.

What do you do with the data you gather from your investigations and expeditions?

Reports of our investigations are published on our website as soon as they are available. Preliminary reports are posted within days of the project finishing.

Each year we publish a 200 page yearbook containing research papers and expedition reports too long to be printed in the journal. We freely circulate our information to anybody who asks for it.

Is the CFZ community purely an electronic one?

No. Each year since 2000 we have held our annual convention - the *Weird Weekend* - in Exeter. It is three days of lectures, workshops, and excursions. But most importantly it is a chance for members of the CFZ to meet each other, and to talk with the members of the permanent directorate in a relaxed and informal setting and preferably with a pint of beer in one hand. Since 2006 - the *Weird Weekend* has been bigger and better and held on the third weekend in August in the idyllic rural location of Woolsery in North Devon.

Since relocating to North Devon in 2005 we have become ever more closely involved with other community organisations, and we hope that this trend will continue. We also work closely with Police Forces across the UK as consultants for animal mutilation cases, and we intend to forge closer links with the coastguard and other community services. We want to work closely with those who regularly travel into the Bristol Channel, so that if the recent trend of exotic animal visitors to our coastal waters continues, we can be out there as soon as possible.

We are building a Visitor's Centre in rural North Devon. This will not be open to the general public, but will provide a museum, a library and an educational resource for our members (currently over 400) across the globe. We are also planning a youth organisation which will involve children and young people in our activities.

Apart from having been the only Fortean Zoological organisation in the world to have consistently published material on all aspects of the subject for over a decade, we have achieved the following concrete results:

- Disproved the myth relating to the headless so-called sea-serpent carcass of Durgan beach in Cornwall 1975
- Disproved the story of the 1988 puma skull of Lustleigh Cleave
- Carried out the only in-depth research ever into the mythos of the Cornish Owlman
- Made the first records of a tropical species of lamprey
- Made the first records of a luminous cave gnat larva in Thailand
- Discovered a possible new species of British mammal - the beech marten
- In 1994-6 carried out the first archival fortean zoological survey of Hong Kong
- In the year 2000, CFZ theories were confirmed when an new species of lizard was added to the British list
- Identified the monster of Martin Mere in Lancashire as a giant wels catfish
- Expanded the known range of Armitage's skink in the Gambia by 80%
- Obtained photographic evidence of the remains of Europe's largest known pike
- Carried out the first ever in-depth study of the *ninki-nanka*
- Carried out the first attempt to breed Puerto Rican cave snails in captivity
- Were the first European explorers to visit the `lost valley` in Sumatra
- Published the first ever evidence for a new tribe of pygmies in Guyana
- Published the first evidence for a new species of caiman in Guyana
- Filmed unknown creatures on a monster-haunted lake in Ireland for the first time
- Had a sighting of orang pendek in Sumatra in 2009
- Published some of the best evidence ever for the almasty in southern Russia

EXPEDITIONS & INVESTIGATIONS TO DATE INCLUDE:

- 1998 Puerto Rico, Florida, Mexico *(Chupacabras)*
- 1999 Nevada *(Bigfoot)*
- 2000 Thailand *(Giant snakes called nagas)*
- 2002 Martin Mere *(Giant catfish)*
- 2002 Cleveland *(Wallaby mutilation)*
- 2003 Bolam Lake *(BHM Reports)*
- 2003 Sumatra *(Orang Pendek)*
- 2003 Texas *(Bigfoot; giant snapping turtles)*
- 2004 Sumatra *(Orang Pendek; cigau, a sabre-toothed cat)*
- 2004 Illinois *(Black panthers; cicada swarm)*
- 2004 Texas *(Mystery blue dog)*
- Loch Morar *(Monster)*
- 2004 Puerto Rico *(Chupacabras; carnivorous cave snails)*
- 2005 Belize *(Affiliate expedition for hairy dwarfs)*
- 2005 Loch Ness *(Monster)*
- 2005 Mongolia *(Allghoi Khorkhoi aka Mongolian death worm)*
- 2006 Gambia *(Gambo - Gambian sea monster, Ninki Nanka and Armitage's skink*
- 2006 Llangorse Lake *(Giant pike, giant eels)*
- 2006 Windermere *(Giant eels)*
- 2007 Coniston Water *(Giant eels)*
- 2007 Guyana *(Giant anaconda, didi, water tiger)*
- 2008 Russia *(Almasty)*
- 2009 Sumatra *(Orang pendek)*
- 2009 Republic of Ireland *(Lake Monster)*
- 2010 Texas *(Blue dogs)*

Other books available from
CFZ PRESS

THE OWLMAN AND OTHERS - 30th Anniversary Edition
Jonathan Downes - ISBN 978-1-905723-02-7

£14.99

EASTER 1976 - Two young girls playing in the churchyard of Mawnan Old Church in southern Cornwall were frightened by what they described as a "nasty bird-man". A series of sightings that has continued to the present day. These grotesque and frightening episodes have fascinated researchers for three decades now, and one man has spent years collecting all the available evidence into a book. To mark the 30th anniversary of these sightings, Jonathan Downes has published a special edition of his book.

DRAGONS - More than a myth?
Richard Freeman - ISBN 0-9512872-9-X

£14.99

First scientific look at dragons since 1884. It looks at dragon legends worldwide, and examines modern sightings of dragon-like creatures, as well as some of the more esoteric theories surrounding dragonkind.

Dragons are discussed from a folkloric, historical and cryptozoological perspective, and Richard Freeman concludes that: "When your parents told you that dragons don't exist - they lied!"

MONSTER HUNTER
Jonathan Downes - ISBN 0-9512872-7-3

£14.99

Jonathan Downes' long-awaited autobiography, *Monster Hunter*...

Written with refreshing candour, it is the extraordinary story of an extraordinary life, in which the author crosses paths with wizards, rock stars, terrorists, and a bewildering array of mythical and not so mythical monsters, and still just about manages to emerge with his sanity intact.......

MONSTER OF THE MERE
Jonathan Downes - ISBN 0-9512872-2-2

£12.50

It all starts on Valentine's Day 2002 when a Lancashire newspaper announces that "Something" has been attacking swans at a nature reserve in Lancashire. Eyewitnesses have reported that a giant unknown creature has been dragging fully grown swans beneath the water at Martin Mere. An intrepid team from the Exeter based Centre for Fortean Zoology, led by the author, make two trips – each of a week – to the lake and its surrounding marshlands. During their investigations they uncover a thrilling and complex web of historical fact and fancy, quasi Fortean occurrences, strange animals and even human sacrifice.

**CFZ PRESS, MYRTLE COTTAGE,
WOOLFARDISWORTHY BIDEFORD,
NORTH DEVON, EX39 5QR
w w w . c f z . o r g . u k**

Other books available from
CFZ PRESS

ONLY FOOLS AND GOATSUCKERS
Jonathan Downes - ISBN 0-9512872-3-0

£12.50

In January and February 1998 Jonathan Downes and Graham Inglis of the Centre for Fortean Zoology spent three and a half weeks in Puerto Rico, Mexico and Florida, accompanied by a film crew from UK Channel 4 TV. Their aim was to make a documentary about the terrifying chupacabra - a vampiric creature that exists somewhere in the grey area between folklore and reality. This remarkable book tells the gripping, sometimes scary, and often hilariously funny story of how the boys from the CFZ did their best to subvert the medium of contemporary TV documentary making and actually do their job.

WHILE THE CAT'S AWAY
Chris Moiser - ISBN: 0-9512872-1-4

£7.99

Over the past thirty years or so there have been numerous sightings of large exotic cats, including black leopards, pumas and lynx, in the South West of England. Former Rhodesian soldier Sam McCall moved to North Devon and became a farmer and pub owner when Rhodesia became Zimbabwe in 1980. Over the years despite many of his pub regulars having seen the "Beast of Exmoor" Sam wasn't at all sure that it existed. Then a series of happenings made him change his mind. Chris Moiser—a zoologist—is well known for his research into the mystery cats of the westcountry. This is his first novel.

CFZ EXPEDITION REPORT 2006 - GAMBIA
ISBN 1905723032

£12.50

In July 2006, The J.T.Downes memorial Gambia Expedition - a six-person team - Chris Moiser, Richard Freeman, Chris Clarke, Oll Lewis, Lisa Dowley and Suzi Marsh went to the Gambia, West Africa. They went in search of a dragon-like creature, known to the natives as `Ninki Nanka`, which has terrorized the tiny African state for generations, and has reportedly killed people as recently as the 1990s. They also went to dig up part of a beach where an amateur naturalist claims to have buried the carcass of a mysterious fifteen foot sea monster named 'Gambo', and they sought to find the Armitage's Skink (*Chalcides armitagei*) - a tiny lizard first described in 1922 and only rediscovered in 1989. Here, for the first time, is their story.... With an forward by Dr. Karl Shuker and introduction by Jonathan Downes.

BIG CATS IN BRITAIN YEARBOOK 2006
Edited by Mark Fraser - ISBN 978-1905723-01-0

£10.00

Big cats are said to roam the British Isles and Ireland even now as you are sitting and reading this. People from all walks of life encounter these mysterious felines on a daily basis in every nook and cranny of these two countries. Most are jet-black, some are white, some are brown, in fact big cats of every description and colour are seen by some unsuspecting person while on his or her daily business. 'Big Cats in Britain' are the largest and most active group in the British Isles and Ireland This is their first book. It contains a run-down of every known big cat sighting in the UK during 2005, together with essays by various luminaries of the British big cat research community which place the phenomenon into scientific, cultural, and historical perspective.

CFZ PRESS, MYRTLE COTTAGE,
WOOLSERY, BIDEFORD,
NORTH DEVON, EX39 5QR
w w w . c f z . o r g . u k

Other books available from
CFZ PRESS

THE SMALLER MYSTERY CARNIVORES OF THE WESTCOUNTRY
Jonathan Downes - ISBN 978-1-905723-05-8

£7.99

Although much has been written in recent years about the mystery big cats which have been reported stalking Westcountry moorlands, little has been written on the subject of the smaller British mystery carnivores. This unique book redresses the balance and examines the current status in the Westcountry of three species thought to be extinct: the Wildcat, the Pine Marten and the Polecat, finding that the truth is far more exciting than the currently held scientific dogma. This book also uncovers evidence suggesting that even more exotic species of small mammal may lurk hitherto unsuspected in the countryside of Devon, Cornwall, Somerset and Dorset.

THE BLACKDOWN MYSTERY
Jonathan Downes - ISBN 978-1-905723-00-3

£7.99

Intrepid members of the CFZ are up to the challenge, and manage to entangle themselves thoroughly in the bizarre trappings of this case. This is the soft underbelly of ufology, rife with unsavoury characters, plenty of drugs and booze." That sums it up quite well, we think. A new edition of the classic 1999 book by legendary fortean author Jonathan Downes. In this remarkable book, Jon weaves a complex tale of conspiracy, anti-conspiracy, quasi-conspiracy and downright lies surrounding an air-crash and alleged UFO incident in Somerset during 1996. However the story is much stranger than that. This excellent and amusing book lifts the lid off much of contemporary forteana and explains far more than it initially promises.

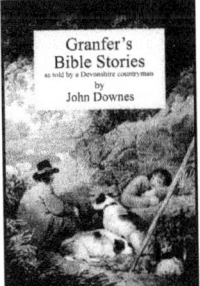

GRANFER'S BIBLE STORIES
John Downes - ISBN 0-9512872-8-1

£7.99

Bible stories in the Devonshire vernacular, each story being told by an old Devon Grandfather - 'Granfer'. These stories are now collected together in a remarkable book presenting selected parts of the Bible as one more-or-less continuous tale in short 'bite sized' stories intended for dipping into or even for bed-time reading. `Granfer` treats the biblical characters as if they were simple country folk living in the next village. Many of the stories are treated with a degree of bucolic humour and kindly irreverence, which not only gives the reader an opportunity to re-evaluate familiar tales in a new light, but do so in both an entertaining and a spiritually uplifting manner.

FRAGRANT HARBOURS DISTANT RIVERS
John Downes - ISBN 0-9512872-5-7

£12.50

Many excellent books have been written about Africa during the second half of the 19th Century, but this one is unique in that it presents the stories of a dozen different people, whose interlinked lives and achievements have as many nuances as any contemporary soap opera. It explains how the events in China and Hong Kong which surrounded the Opium Wars, intimately effected the events in Africa which take up the majority of this book. The author served in the Colonial Service in Nigeria and Hong Kong, during which he found himself following in the footsteps of one of the main characters in this book; Frederick Lugard – the architect of modern Nigeria.

CFZ PRESS, MYRTLE COTTAGE, WOOLFARDISWORTHY BIDEFORD, NORTH DEVON, EX39 5QR
w w w . c f z . o r g . u k

Other books available from
CFZ PRESS

ANIMALS & MEN - Issues 1 - 5 - In the Beginning
Edited by Jonathan Downes - ISBN 0-9512872-6-5

£12.50

At the beginning of the 21st Century monsters still roam the remote, and sometimes not so remote, corners of our planet. It is our job to search for them. The Centre for Fortean Zoology [CFZ] is the only professional, scientific and full-time organisation in the world dedicated to cryptozoology - the study of unknown animals. Since 1992 the CFZ has carried out an unparalleled programme of research and investigation all over the world. We have carried out expeditions to Sumatra (2003 and 2004), Mongolia (2005), Puerto Rico (1998 and 2004), Mexico (1998), Thailand (2000), Florida (1998), Nevada (1999 and 2003), Texas (2003 and 2004), and Illinois (2004). An introductory essay by Jonathan Downes, notes putting each issue into a historical perspective, and a history of the CFZ.

ANIMALS & MEN - Issues 6 - 10 - The Number of the Beast
Edited by Jonathan Downes - ISBN 978-1-905723-06-5

£12.50

At the beginning of the 21st Century monsters still roam the remote, and sometimes not so remote, corners of our planet. It is our job to search for them. The Centre for Fortean Zoology [CFZ] is the only professional, scientific and full-time organisation in the world dedicated to cryptozoology - the study of unknown animals. Since 1992 the CFZ has carried out an unparalleled programme of research and investigation all over the world. We have carried out expeditions to Sumatra (2003 and 2004), Mongolia (2005), Puerto Rico (1998 and 2004), Mexico (1998), Thailand (2000), Florida (1998), Nevada (1999 and 2003), Texas (2003 and 2004), and Illinois (2004). Preface by Mark North and an introductory essay by Jonathan Downes, notes putting each issue into a historical perspective, and a history of the CFZ.

BIG BIRD! Modern Sightings of Flying Monsters

Ken Gerhard - ISBN 978-1-905723-08-9

£7.99

From all over the dusty U.S./Mexican border come hair-raising stories of modern day encounters with winged monsters of immense size and terrifying appearance. Further field sightings of similar creatures are recorded from all around the globe. What lies behind these weird tales? Ken Gerhard is a native Texan, he lives in the homeland of the monster some call 'Big Bird'. Ken's scholarly work is the first of its kind. On the track of the monster, Ken uncovers cases of animal mutilations, attacks on humans and mounting evidence of a stunning zoological discovery ignored by mainstream science. Keep watching the skies!

STRENGTH THROUGH KOI
They saved Hitler's Koi and other stories

Jonathan Downes - ISBN 978-1-905723-04-1

£7.99

Strength through Koi is a book of short stories - some of them true, some of them less so - by noted cryptozoologist and raconteur Jonathan Downes. The stories are all about koi carp, and their interaction with bigfoot, UFOs, and Nazis. Even the late George Harrison makes an appearance. Very funny in parts, this book is highly recommended for anyone with even a passing interest in aquaculture, but should be taken definitely *cum grano salis*.

CFZ PRESS, MYRTLE COTTAGE,
WOOLSERY, BIDEFORD,
NORTH DEVON, EX39 5QR

Other books available from
CFZ PRESS

BIG CATS IN BRITAIN YEARBOOK 2007
Edited by Mark Fraser - ISBN 978-1-905723-09-6

£12.50

People from all walks of life encounter mysterious felids on a daily basis, in every nook and cranny of the UK. Most are jet-black, some are white, some are brown; big cats of every description and colour are seen by some unsuspecting person while on his or her daily business. 'Big Cats in Britain' are the largest and most active research group in the British Isles and Ireland. This book contains a run-down of every known big cat sighting in the UK during 2006, together with essays by various luminaries of the British big cat research community.

CAT FLAPS! Northern Mystery Cats
Andy Roberts - ISBN 978-1-905723-11-9

£6.99

Of all Britain's mystery beasts, the alien big cats are the most renowned. In recent years the notoriety of these uncatchable, out-of-place predators have eclipsed even the Loch Ness Monster. They slink from the shadows to terrorise a community, and then, as often as not, vanish like ghosts. But now film, photographs, livestock kills, and paw prints show that we can no longer deny the existence of these once-legendary beasts. Here then is a case-study, a true lost classic of Fortean research by one of the country's most respected researchers.

CENTRE FOR FORTEAN ZOOLOGY 2007 YEARBOOK
Edited by Jonathan Downes and Richard Freeman
ISBN 978-1-905723-14-0

£12.50

The Centre For Fortean Zoology Yearbook is a collection of papers and essays too long and detailed for publication in the CFZ Journal *Animals & Men*. With contributions from both well-known researchers, and relative newcomers to the field, the Yearbook provides a forum where new theories can be expounded, and work on little-known cryptids discussed.

MONSTER! THE A-Z OF ZOOFORM PHENOMENA
Neil Arnold - ISBN 978-1-905723-10-2

£14.99

Zooform Phenomena are the most elusive, and least understood, mystery `animals`. Indeed, they are not animals at all, and are not even animate in the accepted terms of the word. Author and researcher Neil Arnold is to be commended for a groundbreaking piece of work, and has provided the world's first alphabetical listing of zooforms from around the world.

CFZ PRESS, MYRTLE COTTAGE, WOOLFARDISWORTHY BIDEFORD, NORTH DEVON, EX39 5QR
www.cfz.org.uk

Other books available from
CFZ PRESS

BIG CATS LOOSE IN BRITAIN
Marcus Matthews - ISBN 978-1-905723-12-6

£14.99

Big Cats: Loose in Britain, looks at the body of anecdotal evidence for such creatures: sightings, livestock kills, paw-prints and photographs, and seeks to determine underlying commonalities and threads of evidence. These two strands are repeatedly woven together into a highly readable, yet scientifically compelling, overview of the big cat phenomenon in Britain.

DARK DORSET
TALES OF MYSTERY, WONDER AND TERROR
Robert. J. Newland and Mark. J. North
ISBN 978-1-905723-15-6

£12.50

This extensively illustrated compendium has over 400 tales and references, making this book by far one of the best in its field. Dark Dorset has been thoroughly researched, and includes many new entries and up to date information never before published. The title of the book speaks for itself, and is indeed not for the faint hearted or those easily shocked.

MAN-MONKEY - IN SEARCH OF THE BRITISH BIGFOOT
Nick Redfern - ISBN 978-1-905723-16-4

£9.99

In her 1883 book, *Shropshire Folklore*, Charlotte S. Burne wrote: *'Just before he reached the canal bridge, a strange black creature with great white eyes sprang out of the plantation by the roadside and alighted on his horse's back'*. The creature duly became known as the `Man-Monkey`.

Between 1986 and early 2001, Nick Redfern delved deeply into the mystery of the strange creature of that dark stretch of canal. Now, published for the very first time, are Nick's original interview notes, his files and discoveries; as well as his theories pertaining to what lies at the heart of this diabolical legend.

EXTRAORDINARY ANIMALS REVISITED
Dr Karl Shuker - ISBN 978-1905723171

£14.99

This delightful book is the long-awaited, greatly-expanded new edition of one of Dr Karl Shuker's much-loved early volumes, *Extraordinary Animals Worldwide*. It is a fascinating celebration of what used to be called romantic natural history, examining a dazzling diversity of animal anomalies, creatures of cryptozoology, and all manner of other thought-provoking zoological revelations and continuing controversies down through the ages of wildlife discovery.

CFZ PRESS, MYRTLE COTTAGE, WOOLFARDISWORTHY BIDEFORD, NORTH DEVON, EX39 5QR
www.cfz.org.uk

Other books available from
CFZ PRESS

DARK DORSET CALENDAR CUSTOMS
Robert J Newland - ISBN 978-1-905723-18-8

£12.50

Much of the intrinsic charm of Dorset folklore is owed to the importance of folk customs. Today only a small amount of these curious and occasionally eccentric customs have survived, while those that still continue have, for many of us, lost their original significance. Why do we eat pancakes on Shrove Tuesday? Why do children dance around the maypole on May Day? Why do we carve pumpkin lanterns at Hallowe'en? All the answers are here! Robert has made an in-depth study of the Dorset country calendar identifying the major feast-days, holidays and celebrations when traditionally such folk customs are practiced.

CENTRE FOR FORTEAN ZOOLOGY 2004 YEARBOOK
Edited by Jonathan Downes and Richard Freeman
ISBN 978-1-905723-14-0

£12.50

The Centre For Fortean Zoology Yearbook is a collection of papers and essays too long and detailed for publication in the CFZ Journal *Animals & Men*. With contributions from both well-known researchers, and relative newcomers to the field, the Yearbook provides a forum where new theories can be expounded, and work on little-known cryptids discussed.

CENTRE FOR FORTEAN ZOOLOGY 2008 YEARBOOK
Edited by Jonathan Downes and Corinna Downes
ISBN 978 -1-905723-19-5

£12.50

The Centre For Fortean Zoology Yearbook is a collection of papers and essays too long and detailed for publication in the CFZ Journal *Animals & Men*. With contributions from both well-known researchers, and relative newcomers to the field, the Yearbook provides a forum where new theories can be expounded, and work on little-known cryptids discussed.

ETHNA'S JOURNAL
Corinna Newton Downes
ISBN 978 -1-905723-21-8

£9.99

Ethna's Journal tells the story of a few months in an alternate Dark Ages, seen through the eyes of Ethna, daughter of Lord Edric. She is an unsophisticated girl from the fortress town of Cragnuth, somewhere in the north of England, who reluctantly gets embroiled in a web of treachery, sorcery and bloody war...

**CFZ PRESS, MYRTLE COTTAGE,
WOOLFARDISWORTHY BIDEFORD,
NORTH DEVON, EX39 5QR
www.cfz.org.uk**

Other books available from
CFZ PRESS

ANIMALS & MEN - Issues 11 - 15 - The Call of the Wild
Jonathan Downes (Ed) - ISBN 978-1-905723-07-2

£12.50

Since 1994 we have been publishing the world's only dedicated cryptozoology magazine, *Animals & Men*. This volume contains fascimile reprints of issues 11 to 15 and includes articles covering out of place walruses, feathered dinosaurs, possible North American ground sloth survival, the theory of initial bipedalism, mystery whales, mitten crabs in Britain, Barbary lions, out of place animals in Germany, mystery pangolins, the barking beast of Bath, Yorkshire ABCs, Molly the singing oyster, singing mice, the dragons of Yorkshire, singing mice, the bigfoot murders, waspman, British beavers, the migo, Nessie, the weird warbling whatsit of the westcountry, the quagga project and much more...

IN THE WAKE OF BERNARD HEUVELMANS
Michael A Woodley - ISBN 978-1-905723-20-1

£9.99

Everyone is familiar with the nautical maps from the middle ages that were liberally festooned with images of exotic and monstrous animals, but the truth of the matter is that the *idea* of the sea monster is probably as old as humankind itself.

For two hundred years, scientists have been producing speculative classifications of sea serpents, attempting to place them within a zoological framework. This book looks at these successive classification models, and using a new formula produces a sea serpent classification for the 21st Century.

CENTRE FOR FORTEAN ZOOLOGY 1999 YEARBOOK
Edited by Jonathan Downes
ISBN 978 -1-905723-24-9

£12.50

The Centre For Fortean Zoology Yearbook is a collection of papers and essays too long and detailed for publication in the CFZ Journal *Animals & Men*. With contributions from both well-known researchers, and relative newcomers to the field, the Yearbook provides a forum where new theories can be expounded, and work on little-known cryptids discussed.

CENTRE FOR FORTEAN ZOOLOGY 1996 YEARBOOK
Edited by Jonathan Downes
ISBN 978 -1-905723-22-5

£12.50

The Centre For Fortean Zoology Yearbook is a collection of papers and essays too long and detailed for publication in the CFZ Journal *Animals & Men*. With contributions from both well-known researchers, and relative newcomers to the field, the Yearbook provides a forum where new theories can be expounded, and work on little-known cryptids discussed.

**CFZ PRESS, MYRTLE COTTAGE,
WOOLFARDISWORTHY BIDEFORD,
NORTH DEVON, EX39 5QR
www.cfz.org.uk**

Other books available from
CFZ PRESS

BIG CATS IN BRITAIN YEARBOOK 2008
Edited by Mark Fraser - ISBN 978-1-905723-23-2

£12.50

People from all walks of life encounter mysterious felids on a daily basis, in every nook and cranny of the UK. Most are jet-black, some are white, some are brown; big cats of every description and colour are seen by some unsuspecting person while on his or her daily business. 'Big Cats in Britain' are the largest and most active research group in the British Isles and Ireland. This book contains a run-down of every known big cat sighting in the UK during 2007, together with essays by various luminaries of the British big cat research community.

CFZ EXPEDITION REPORT 2007 - GUYANA
ISBN 978-1-905723-25-6

£12.50

Since 1992, the CFZ has carried out an unparalleled programme of research and investigation all over the world. In November 2007, a five-person team - Richard Freeman, Chris Clarke, Paul Rose, Lisa Dowley and Jon Hare went to Guyana, South America. They went in search of giant anacondas, the bigfoot-like didi, and the terrifying water tiger.

Here, for the first time, is their story...With an introduction by Jonathan Downes and forward by Dr. Karl Shuker.

CENTRE FOR FORTEAN ZOOLOGY 2003 YEARBOOK
Edited by Jonathan Downes and Richard Freeman
ISBN 978-1-905723-19-5

£12.50

The Centre For Fortean Zoology Yearbook is a collection of papers and essays too long and detailed for publication in the CFZ Journal *Animals & Men*. With contributions from both well-known researchers, and relative newcomers to the field, the Yearbook provides a forum where new theories can be expounded, and work on little-known cryptids discussed.

CENTRE FOR FORTEAN ZOOLOGY 1997 YEARBOOK
Edited by Jonathan Downes and Graham Inglis
ISBN 978-1-905723-27-0

£12.50

The Centre For Fortean Zoology Yearbook is a collection of papers and essays too long and detailed for publication in the CFZ Journal *Animals & Men*. With contributions from both well-known researchers, and relative newcomers to the field, the Yearbook provides a forum where new theories can be expounded, and work on little-known cryptids discussed.

**CFZ PRESS, MYRTLE COTTAGE,
WOOLFARDISWORTHY BIDEFORD,
NORTH DEVON, EX39 5QR
www.cfz.org.uk**

Other books available from
CFZ PRESS

CENTRE FOR FORTEAN ZOOLOGY 2000-1 YEARBOOK
Edited by Jonathan Downes and Richard Freeman
ISBN 978-1-905723-19-5

£12.50

The Centre For Fortean Zoology Yearbook is a collection of papers and essays too long and detailed for publication in the CFZ Journal *Animals & Men*. With contributions from both well-known researchers, and relative newcomers to the field, the Yearbook provides a forum where new theories can be expounded, and work on little-known cryptids discussed.

**THE MYSTERY ANIMALS OF THE BRITISH ISLES:
NORTHUMBERLAND AND TYNESIDE**
Michael J Hallowell
ISBN 978-1-905723-29-4

£12.50

Mystery animals? Great Britain? Surely not. But is is true.

This is a major new series from CFZ Press. It will cover Great Britain and the Republic of Ireland, on a county by county basis, describing the mystery animals of the entire island group.

The Island of Paradise: Chupacabra, UFO Crash Retrievals, and Accelerated Evolution on the Island of Puerto Rico
Jonathan Downes - ISBN 978-1-905723-32-4

£14.99

In his first book of original research for four years, Jon Downes visits the Antillean island of Puerto Rico, to which he has led two expeditions - in 1998 and 2004. Together with noted researcher Nick Redfern he goes in search of the grotesque vampiric chupacabra, believing that it can - finally - be categorised within a zoological frame of reference rather than a purely paranormal one. Along the way he uncovers mystery after mystery, has a run in with terrorists, art historians, and even has his garden buzzed by a UFO. By turns both terrifying and funny, this remarkable book is a real tour de force by one of the world's foremost cryptozoological researchers.

DR SHUKER'S CASEBOOK
Dr Karl Shuker - ISBN 978-1905723-33-1

£14.99

Although he is best-known for his extensive cryptozoological researches and publications, Dr Karl Shuker has also investigated a very diverse range of other anomalies and unexplained phenomena, both in the literature and in the field. Now, compiled here for the very first time, are some of the extraordinary cases that he has re-examined or personally explored down through the years.

CFZ PRESS, MYRTLE COTTAGE,
WOOLFARDISWORTHY BIDEFORD,
NORTH DEVON, EX39 5QR
www.cfz.org.uk

Other books available from
CFZ PRESS

Dinosaurs and Other Prehistoric Animals on Stamps: A Worldwide Catalogue
Dr Karl P.N.Shuker - ISBN 978-1-905723-34-8

£9.99

Compiled by zoologist Dr Karl P.N. Shuker, a lifelong, enthusiastic collector of wildlife stamps and with an especial interest in those that portray fossil species, it provides an exhaustive, definitive listing of stamps and miniature sheets depicting dinosaurs and other prehistoric animals issued by countries throughout the world. It also includes sections dealing with cryptozoological stamps, dinosaur stamp superlatives, and unofficial prehistoric animal stamps.

CFZ EXPEDITION REPORT 2008 - RUSSIA
ISBN 978-1-905723-35-5

Since 1992, the CFZ has carried out an unparalleled programme of research and investigation all over the world. In July 2008, a five-person team - Richard Freeman, Chris Clarke, Dave Archer, Adam Davies and Keith Townley went to Kabardino-Balkaria in southern Russia in search of the almasty, maybe mankind's closest relative. Here, for the first time, is their story...With an introduction by Jonathan Downes and forward by Dr. Karl Shuker.

CENTRE FOR FORTEAN ZOOLOGY 2009 YEARBOOK
Edited by Jonathan Downes and Richard Freeman
ISBN 978 -1-905723-37

£12.50

The Centre For Fortean Zoology Yearbook is a collection of papers and essays too long and detailed for publication in the CFZ Journal *Animals & Men*. With contributions from both well-known researchers, and relative newcomers to the field, the Yearbook provides a forum where new theories can be expounded, and work on little-known cryptids discussed.

THE MYSTERY ANIMALS OF THE BRITISH ISLES: KENT
Neil Arnold
ISBN 978-1-905723-36-2

£12.50

Mystery animals? Great Britain? Surely not. But is is true.

This is a major new series from CFZ Press. It will cover Great Britain and the Republic of Ireland, on a county by county basis, describing the mystery animals of the entire island group.

**CFZ PRESS, MYRTLE COTTAGE,
WOOLFARDISWORTHY BIDEFORD,
NORTH DEVON, EX39 5QR
www.cfz.org.uk**

Other books available from
CFZ PRESS

GIANT SNAKES
By Michael Newton
ISBN: 978-1-905723-39-3

£9.99

In this exciting book, Michael Newton takes an overview of the most terrifying uberpredators in the world - giant snakes. Outsized examples of known species as well as putative new species are looked at in detail.

THE MYSTERY ANIMALS OF THE BRITISH ISLES: THE WESTERN ISLES
Glen Vaudrey
ISBN 978-1-905723-42-3

£12.50

Mystery animals? Great Britain? Surely not. But is is true.

This is a major new series from CFZ Press. It will cover Great Britain and the Republic of Ireland, on a county by county basis, describing the mystery animals of the entire island group.

Strangely Strange but Oddly Normal
Andy Roberts
ISBN 978-1-905723-44-7

£11.99

An anthology of writings from one of Britain's most respected Fortean authors, covering everything from UFOs, to the Rolling Stones, and from psychedelic drugs to ancient fertility symbols, the Incredible String Band, and government cover-ups.

China: The Yellow Peril?
Richard Muirhead
ISBN 978-1-905723-41-6

£7.99

Richard Muirhead takes an in depth look at the history of Western relationships with China. If some Victorian antiquarians are to be believed contact between the Chinese Empire and other Middle Eastern and Western Empires goes back to times long before the birth of Christ, such as the ancient Egyptians and the Roman Empire.

**CFZ PRESS, MYRTLE COTTAGE,
WOOLFARDISWORTHY BIDEFORD,
NORTH DEVON, EX39 5QR
www.cfz.org.uk**

Other books available from
CFZ PRESS

CFZ PRESS

CENTRE FOR FORTEAN ZOOLOGY 2009 YEARBOOK
Edited by Jonathan Downes and Richard Freeman
ISBN 978-1-905723-37

£12.50

edited by
Jonathan and Corinna
Downes

The Centre For Fortean Zoology Yearbook is a collection of papers and essays too long and detailed for publication in the CFZ Journal *Animals & Men*. With contributions from both well-known researchers, and relative newcomers to the field, the Yearbook provides a forum where new theories can be expounded, and work on little-known cryptids discussed.

A DAINTREE DIARY
By Carl Portman
ISBN: 978-1-905723-53-9

£9.99

Carl Portman, tarantula expert from the West Midlands travels to Queensland with two Sheilas in search of spiders and adventure. He finds both in this engaging and oddly heartwarming book. A must for anyone interested in the natural world. A glorious mix of arachnology and adventure, with a smattering of silly humour. Not to be missed

PREDATOR DEATHMATCH
By Nick Molloy
ISBN: 978-1-905723-45-4

£8.99

Predator Deathmatch is the first ever book to study apex predators and actually pose the question of who is/was the ultimate predator by pitting them against each other. The author has carefully profiled each contender with a mixture of historical data, information from the fossil record and current observations of wild animal behaviour. .

**CFZ PRESS, MYRTLE COTTAGE,
WOOLFARDISWORTHY BIDEFORD,
NORTH DEVON, EX39 5QR
www.cfz.org.uk**

Reader's Notes

www.ingramcontent.com/pod-product-compliance
Lightning Source LLC
Chambersburg PA
CBHW051051160426
43193CB00010B/1148